Scale and Access Issues Affecting Smallholder Hog Producers in an Expanding Peri-Urban Market

Southern Luzon, Philippines

Achilles Costales
Christopher Delgado
Maria Angeles Catelo
Ma. Lucila Lapar
Marites Tiongco
Simeon Ehui
and
Anne Zillah Bautista

RESEARCH REPORT 151

INTERNATIONAL FOOD POLICY RESEARCH INSTITUTE
sustainable solutions for ending hunger and poverty
IFPRI

University of the Philippines Los Baños

ILRI
INTERNATIONAL LIVESTOCK RESEARCH INSTITUTE

CGIAR Systemwide Livestock Programme

International Food Policy Research Institute
2033 K Street, NW
Washington, D.C. 20006-1002, U.S.A.
Telephone +1-202-862-5600
www.ifpri.org

DOI: 10.2499/9780896291591RR151

Library of Congress Cataloging-in-Publication Data

Scale and access issues affecting smallholder hog producers in an expanding peri-urban market : southern Luzon, Philippines / Achilles Costales . . . [et al.].
p. cm. — (Research report ; 151)
Includes bibliographical references.
ISBN-13: 978-0-89629-159-1 (alk. paper)
ISBN-10: 0-89629-159-6 (alk. paper)
1. Pork industry and trade—Philippines. 2. Swine breeders—Philippines. 3. Farms, small—Philippines. 4. Agriculture and state—Philippines. I. Costales, Achilles. II. Series: Research report (International Food Policy Research Institute) ; 151.
HD9435.P612S25 2006
338.1'7640095991—dc22 2006033435

Contents

Tables

Figures

Foreword

Research has shown that gainful participation in livestock markets is an important means of reducing poverty in developing countries, particularly for rural and peri-urban households. The rapid growth in demand for meat and milk, along with the corresponding expansion of livestock markets to connect consumers and suppliers, presents real opportunities for smallholders to generate income by raising livestock. Nevertheless, the structural changes associated with increasing urbanization taking place in these markets, the greater integration between domestic and global markets, and the emergence of a more stringent regulatory environment also present significant threats to participation by poor households. Further, as the market for livestock products rapidly grows, smallholders have to compete with large-scale commercial producers for market share, particularly at the domestic level, and if market forces and policy environments are biased toward larger-scale producers, smallholders are often displaced. IFPRI has identified research on the future of smallholder farming as a priority for improving our understanding of the relationship between livestock sector development and poverty reduction, thereby enhancing opportunities for smallholders, and countering threats. To this end, this study begins by examining the market forces, structural factors, and policies that affect the scale of pig production, and then goes on to identify strategies for enhancing smallholder participation and competitiveness in a rapidly growing livestock market.

The study offers a new way of conceptualizing the problems that lead to the exclusion of smallholders from live hog and pork markets, explaining why some smallholders participate successfully, while others do not. Determinants are identified using limited-dependent variable models based on the hypothesis that transaction costs, such as access to credit and market information, affect market participation. The report also presents a contemporary approach to measuring profit efficiency in hog production for the case of Southern Luzon, Philippines.

Although the findings of this study are specific to the Philippine context, many of the issues confronted are common to the challenges of participation, upscaling processes, and policy interventions across the developing world. The research has generated solid empirical perspectives of the changing situation of poor smallholder producers in a high-value market situation. IFPRI thus continues to examine the effect of mechanisms like contract farming on collective action as a means of increasing smallholder participation in high-value markets, particularly in developing countries in Asia and Africa, where small farms continue to dominate the landscape.

Joachim von Braun
Director General, IFPRI

Acknowledgments

This study would not have been possible without the kindness of individuals in the field who assisted the authors in the preparation and undertaking of the field survey of smallholder and commercial hog producers, of colleagues who lent their sharp minds, and of institutions that provided logistical support for the study. The authors extend special acknowledgments to the following individuals and institutions, with the order of mention not necessarily according to importance:

The Systemwide Livestock Programme (SLP) of the Consultative Group on International Agricultural Research (CGIAR) centers—especially the intercenter Livestock Programme Group—and its sponsors, for including the study as part of the SLP research agenda, providing the financial support, and making this study the first of its kind in the Philippines;

All our respondents—participants and nonparticipants in the livestock activity—but especially to smallholder hog producers, who gave us a rare and rich insight into the world of smallholders, coping with the odds of competition in an exciting market whose playing field is often uneven;

Dr. Leoncio Rebong Jr., Laguna Provincial veterinarian, who had encouraged the team to undertake more studies of the hog business in his province;

Loreto Sayson, San Pablo City agriculturist, and Dr. Sam Comia of the Office of the City Veterinarian of Lipa City, for providing a list of sample *barangays* (villages) and prospective sample smallholders, and for providing staff to accompany the team and enumerators to the individual households identified for interviews;

Romy Villarasa of the Municipal Agricultural Office of Majayjay, for accompanying and introducing the team to smallholders and commercial pig producers in Majayjay, Laguna, during the pretesting of the instruments;

Rico Geron, general manager of the Sorosoro Ibaba Development Cooperative (SIDC), for engaging us with vibrant discussions on the hog contract growing (*paiwi*) program of the cooperative;

Officers and staff of the SIDC, Marifie Dimaano and Andrea Villa of the Main Office, for providing us with the documentary information on the Cooperative, and for patiently being there, smiles undiminished, to answer follow-up questions on how the Cooperative and the hog contract growing program worked; Analita Bautro, livestock manager, for providing staff to accompany the enumerators to the sample contract growers identified; Ruben Coliyat, marketing head, for giving vital information on selection, pricing, and distribution of growing stocks; and Lucido Plata, feedmill manager, for showing that quality feeds for smallholders can be produced, but that much depends on the prices of feedgrains;

Dr. Valentino Argañosa, for his wise advice on relevant technical parameters in commercial hog enterprises and for personally endorsing the team to commercial hog producers for a chance to be entertained for interviews on quite sensitive business information;

Raymond Galamgam, of the Philippine Council for Agriculture, Forestry, and Natural Resources Research and Development–Philippine Swine Industry Development Foundation,

Inc., for assisting in the interviews with commercial hog producers who were themselves quite difficult to catch;

Dr. Elpidio Agbisit Jr., of the Institute of Animal Science, University of the Philippines Los Baños, for helping the team with technical parameters in smallholder hog raising, to countercheck whether the production information provided by the respondents were within reasonable bounds;

Dr. Cristina David, Dr. Agnes Rola, and Dr. James Mateo, for their incisive comments on the study during the presentation of the results in the Systemwide Livestock Program Outreach Workshop in November 2002;

Catherine Aragon, Amelia delos Reyes and Amal Chatterjee Jr., for taking up the cudgels for us through thick and thin;

The International Livestock Research Institute office staff for administrative assistance;

Dr. Douglas Gray for his unstinting support and encouragement in his capacity as the International Livestock Research Institute regional representative for Southeast Asia;

Dr. Romeo Recide, director of the Bureau of Agricultural Statistics and assistant secretary for Policy and Planning of the Department of Agriculture, for his invaluable assistance for letting us access the registry of commercial pig farms and size distribution of capacity, by province and by region for 2004; and Nenita Yanson, chief of the Livestock Section, Livestock and Poultry Statistics Division of the Bureau of Agricultural Statistics, for providing us with the latest inventory of backyard and commercial pig inventories in the Philippines; Central Luzon; Southern Tagalog; and the Cavite, Laguna, Batangas, Rizal, and Quezon subregion of Southern Tagalog;

The UPLB Foundation, Inc., for giving institutional legitimacy to the project within the complex of rules for researches done in the university; and the Department of Economics, for providing invaluable space for the team to work, even at unholy hours of the morning.

Summary

Among all sectors in the Philippine agricultural economy, livestock exhibited the fastest and most consistent growth (4.6 percent on average per year) in the 1990s and well into the twenty-first century, steadily increasing its share of contribution to gross value added in agriculture from 18 percent in 1990 to 24 percent by 2003. The expansion of the industry has been propelled mainly by growth in domestic demand for meat in general, and for pork in particular, fueled by a still-rapid population growth, increased urbanization, and modest improvements in per capita income, particularly in the national capital and in major urban centers in the provinces around the Metropolitan Manila area.

Pig production is the largest contributor to meat output in the Philippines. It is also an economic activity in which smallholders still dominate, accounting for close to 80 percent of total pig inventories. The strong growth in demand for pork presents a potential for increasing income opportunities, and therefore for poverty alleviation among rural and agricultural households in the Philippines, where rural poverty incidence for families remains high at 40.3 percent. In the two major hog-producing regions of Southern and Central Luzon adjacent to the national capital, however, the observed trends depict a more rapid decline in the share of smallholder pig producers in regional output, such that by 2003, smallholders held the minority share. In these two regions, although the number of registered commercial pig farms significantly increased, the number of farms raising pigs declined between the 1990s and 2000s. These numbers suggest a scaling up of larger farms and a displacement of smaller ones.

The more rapid growth in larger commercial farms is not solely premised on the existence of economies of scale in production but is attributed, in large part, to the commercial sector being able to escape or at least overcome transaction costs that cannot be handled as easily by small producers. In addition, the cost advantage of large farms over small ones is also hypothesized to stem from their ability to access privileged prices of crucial inputs that are linked to policy subsidies.

On the demand side of the meat industry, consuming households in the metropolis and in the major urbanizing centers in the two regions, with their higher incomes and purchasing power, are not only increasing their consumption of meat but also their demand for quality, convenience, and greater product differentiation. On the supply side, large meat products companies are vertically integrating pig-production operations; exercising control over their own breeding farms, feed formulation, and animal health services; and establishing Hazard Analysis and Critical Control Point–compatible slaughtering and processing plants. They are also producing differentiated meat cuts and meat products, packaging them in convenient forms, and establishing their own brands. In the regulatory scene, the government's National Meat Inspection Service is strengthening monitoring, inspection, and grading functions on the sanitary conditions of all slaughterhouses, public and private, for local, national, and international trade in meat. These changing market forces, compounded by structural factors and the policy environment, pose a serious challenge to the continued survival of smallholder pig producers.

The poverty implications of the displacement of smallholders in the market raise important social questions. This report attempts to respond to these questions by investigating empirical evidence about the scale and access issues that affect smallholders. The study employs two methods to quantify the contributions of market forces and policies affecting the scale of livestock-production operations. Then it relates each contribution to measures that can be taken to overcome transaction cost barriers and policy distortions faced by smallholders that tend to ease them out of mainstream markets for their output and consign them to marginal ones.

Chapter 1 provides the background of the study, states the objectives, and lays down the scale issues related to competitiveness in markets for livestock that are to be investigated. The chapter breaks down the objectives into three main questions on: (1) whether small pig-producing farms can compete with large farms; (2) the role of transaction cost barriers in smallholder participation and performance in hog production; and (3) redressing transaction cost barriers if these barriers affect small and large farms differentially.

Chapter 2 lays out the two approaches that the study uses in investigating the role of transaction costs in economic activities. The first is a direct approach through a market participation model. The second is an indirect approach through the estimation of an efficiency model. A brief review of the literature on these approaches is provided. The market participation model assesses the roles of household characteristics and factors outside the household in determining participation in pig production and the level of market participation. The profit-efficiency model assesses farm-level differences in efficiency in generating profits from a given level of resources, input and output prices in terms of differences in transaction costs, and policy subsidies across the scale of operations studied (from small, backyard-level to medium-scale farms).

Chapter 3 focuses on the determinants of participation by households in the economic activity of pig production. It begins by providing a brief description of the Southern Tagalog region and the subregion consisting of Cavite, Laguna, Batangas, Rizal, and Quezon provinces (CALABARZON), where the study was undertaken. The investigation draws on field data collected in 2000–2001 from three cities in the two major pig-producing provinces in the region. Household-level data were obtained from a sample of 144 pig-producing households and 141 households not producing livestock, randomly picked from identified pig-producing villages.[1] A probit market-participation model was estimated and results were analyzed to explain why some households engage in pig production whereas others do not. Among the significant factors that influenced the decision of households to participate in pig production were the availability of household resources, particularly family labor, and the opportunity costs of the household head and spouse in engaging in pig production, a labor-intensive economic activity. Households with more members of working age tended to participate, whereas those with household heads or spouses that were government or private sector employees tended not to be engaged in pig production. The decision to participate was also influenced by the capacity of the households to deal with fixed transaction costs related to access to financial resources for engaging in the activity and to negotiations involved in a market-oriented activity. The results of the Heckman two-step selection model reveal that the price of slaughter hogs and the access to market outlets that are not limited to the confines of the village are significant factors that influenced the household's level of participation in pig production.

Chapter 4 focuses on the pig-producing households, further disaggregated into independent and contract producers. The chapter compares independent and contract farms in terms of the types of production activities engaged in and access to information and credit, feeds and growing or breeding stock, animal health services, and markets for inputs and outputs.

[1]Fifteen out of 300 households were lost because of sample attrition.

Most contract growers tended to specialize in fattening pigs to slaughter hogs. In contrast, independent producers tended to combine the production of weaners (piglets) with slaughter hogs or specialized in weaner piglet production. Both independent and contract producers were highly market oriented, although in varying intensities, from purchasing mixed feeds and growing or breeding stock, to selling output to various outlets.

Smallholders were grouped into size quintiles according to the level of their annualized outputs. The levels of activity covered a wide range. Those belonging to the lowest quintile produced very small volumes of output (averaging 315 kilograms per year). Mean output was equivalent to approximately five slaughter hogs per year, or about 20 piglets per year, in one or two batches. Producers in the third quintile produced about six times as much as the first, whereas those in the last quintile produced about 200 slaughter hogs per year, 40 times more than the average production per farm of the first quintile.

In general, contract growers in the sample operated at significantly higher levels of activity than did independents, with the majority of them falling in the fourth and fifth size quintiles. Even so, they are still rightfully classified as smallholder producers engaged in a nonformal enterprise, having very similar household characteristics to the independent smallholder sample except for the number of animals kept. The two groups mainly use unpaid household labor (rarely employing hired labor) and have similar nonlivestock sector resources. Contract growers exhibited better access to quality feeds and stock, feed credit, veterinary health services, and credit for expansion purposes. Contract growers also had better access to markets for slaughter hogs, even if their location relative to that of independent producers was at a greater distance from Metropolitan Manila, the main market for live hogs. The advantage of contract growers in the market relative to independents can be traced to the former's integration into a formal institution—a feedmilling and multipurpose cooperative—for the production and marketing of their output. Because of the interesting and innovative features of the contract growing scheme, which helped smallholder pig producers overcome transaction cost barriers through access of information, technology, and markets, an Appendix to this report describes in greater detail the institutional arrangements that smallholders had with their multipurpose cooperative.

Chapter 5 compares farm profits per unit of output of smallholder pig producers, according to scale of production, grouped *ex post* into quintiles. Smallholders specializing in the production of weaners were excluded from the quintile grouping and treated as a separate sample. Profits were computed with and without imputing the costs of family labor at market wage rates. The results showed that when family labor was not costed, the group of pig producers in lowest quintile had the lowest profit performance. Progressing from the smallest to the largest farms, profit performance sharply rises from the first to the second quintile, gradually falls to the fourth quintile, and then moves up again at the last quintile. When family labor was imputed a cost using the minimum rate for agricultural wage workers, the smallholders in the lowest quintile were the worst performers, with their profits almost disappearing. The profit performance follows a similar pattern when family labor was not costed. Thus, smallholders making up the first quintile would not likely survive under conditions of increasing competition. The smallholders in the next two quintiles of smaller farms performed relatively well, even managing to post nominally higher profit per unit of output than did larger farms when family labor was not costed. Even when family labor was costed, these groups of smaller farms still, on average, registered profits per unit of output that were comparable to those earned by larger farms. This group of smaller farms have a decent chance of competing with larger farms.

The profit performance of independent farms was also compared with that of contract growers. When no cost is put on family labor, there was no significant difference in the mean performance between the two groups. When labor cost was imputed, however, independent producers performed significantly worse on average than did contract growers. The cost of family labor thus mattered more to the relative competitiveness of independent producers than to that of contract growers. This result mainly stemmed from the larger scale of operations of contract growers compared to that of independents.

Chapter 5 also presents the estimates of the profit efficiency of smallholder farms and the factors contributing to their inefficiency. In general, the pattern of relative efficiency followed the pattern of profit per unit of output for the quintiles. The smallest farms were the least efficient in generating profits, given their resources and prices of inputs and outputs. This group of producers is not in a position to compete in the market.

The frontier estimation results presented in Chapter 5 showed that the prices of slaughter hogs, feeds, and growing stock were significant determinants of profit performance. Profit per unit of output was most sensitive to the price of feed. Access to reliable feed was the most crucial factor in reducing inefficiency, followed by the existence of other income sources of the household head. When the sample of smallholder producers was bisected by farm size to explain the differences in profit inefficiencies, transaction cost barriers linked to the access to feeds of known quality and access to veterinary services had significant effects on the efficiency of the group of smaller farms, but not to the group of larger producers.

The last chapter of the report (Chapter 6) is devoted to the policy implications derived from the results of the study. It summarizes the conditions under which smallholders can reasonably be expected to compete with larger, more commercially oriented farms for some time, at least under the current level and nature of market demand.[2] The study concludes that the segment of smallholder producers with the smallest holdings of pigs (fewer than 10 slaughter hogs per year) will likely not survive market competition for much longer, by virtue of their relative profit inefficiency and their very low profit performance relative to the rest. Furthermore, their very low levels of output will not allow them to continue to make a living from this activity. Exit for this group may not be immediate but is likely to be inevitable. Public policy in this case might focus on alternative schemes for poverty alleviation for this group, particularly those directed at improving the human capital of households to allow them to be more easily absorbed in lucrative nonfarm employment, as well as toward improving the environment in which business enterprises can flourish.

Apart from the group with very low levels of operation, many smallholders, even those with quite modest levels of output (20–40 slaughter hogs per year) are quite competitive: they can be as efficient as the larger farms in earning profits from pig production, even when the opportunity cost of family labor is considered. Even so, the smaller producers bear with greater difficulty the adverse effects of transaction costs barriers than do larger producers.

Participation in market-oriented livestock production is more likely by households with lower opportunity costs of labor, but barriers exist for those that have little access to capital and financial resources to start the activity. Policy interventions to enhance participation among such households could be directed at improving the business investment climate and institutional environment for commercial enterprises with resources and technological expertise in livestock production (nutrition, breeding, and animal health). These commercial enterprises would then find it profitable to invest in smallholder producers, financing the intermediate input requirements, while taking advantage of the willingness of households to engage in livestock production at wages that are below legislated or market wages.

[2]The study cannot evaluate the impact of a shift in the nature of demand, such as the rise in demand for leaner or safety-certified pork, both of which seem likely at some point, but which are presently not major factors in the Metro Manila market.

This study found that access to feeds of known quality and access to veterinary services were critical transaction cost barriers to smaller producers but not to larger farms in determining profit efficiency and thus competitiveness. Hence, a policy addressing these issues can improve the performance and viability of smallholder pig producers in general. To improve information on feed quality, government can devise simple and clear rules on feed and feed ingredient labeling, accompanied by straightforward methods of spot-checking and proper enforcement. This practice can be combined with measures that facilitate vertical coordination schemes between firms engaged in commercial feed and/or livestock production for higher-value markets and smallholders with pig-raising expertise and facilities.

Among the variables hypothesized to affect profit efficiency, the price of feed has the strongest significant effect on profit performance of the smallholder producers. Thus feed pricing policy would have an important impact on the competitiveness of pig production in general and the viability of smallholder pig production in particular. At present, the domestic feedgrains industry is protected by relatively high tariff rates from import competition (35 percent for in-quota, and 50 percent for out-quota import levels), aimed at protecting domestically produced corn. The livelihoods of smallholder farmers throughout the country, of which more than 40 percent are raising pigs, are adversely affected by a feedgrains policy distortion that is purportedly pro-poor. Hence, removal of such protective restrictions on feedgrains will undoubtedly boost the livestock sector in general. Finally, this policy needs to be complemented with cost-effective measures in producing corn at the farm level and improved transport and distribution infrastructures at the marketing level to improve efficiency in the domestic corn industry.

CHAPTER 1

Introduction

This chapter provides the background and rationale of the study. It states the scale issues related to competitiveness in markets for livestock and poses three research questions relevant to those issues that the study investigates formally.

Changing Trends in Livestock and Farming

Among all sectors in Philippine agriculture, livestock exhibited the fastest growth over the past decade, posting an average annual growth of 4.6 percent in real terms between 1990 and 2003, compared to 2.4 percent in fisheries, 1.6 percent in crops, and a decline (–15 percent) in forestry. The livestock sector steadily increased its share of gross value added in agriculture from 18 percent in 1990 to 24 percent by 2003 (NSCB 2005; NSO 2005). Livestock output steadily expanded at a rate of 5.8 percent per year over the same period (FAO 2005a). The main activity in livestock is pig production, accounting for about 58 percent of total meat output and growing at 5.5 percent per year.

Growth in the livestock sector has been mainly driven by growth in domestic demand, fueled by a still-rapid pace of population growth (2.4 percent per year between 1995 and 2000), increased urbanization, and modest improvements in per capita incomes from the second half of the 1990s through 2003 (ADB 2005; NSCB 2005; NSO 2005). Estimated per capita consumption of meat, derived from the Food and Agriculture Organization of the United Nations (FAO) Food Balance Sheets (where per capita food supply includes net imports) rose from 18 kilograms in 1990 to 31 kilograms by 2002 (FAO 2005b). In the past decade, growth in domestic demand of meat was increasingly supplemented by imports, increasing at the high rate of 25 percent per year during 1990–2002. As imports had started from a very low base value, the share of imports in total supply remained less than 5 percent in 2002 (FAO 2005a).

The rapid growth in domestic demand for pork presents a potential for increased smallholder income and thus for poverty alleviation among rural households in the Philippines, where poverty incidence remains high at 40.3 percent (NSCB 2005).[3] Pig production is an economic activity that is still dominated by smallholders. Close to 80 percent of total hog inventories is held in what the Philippines Bureau of Agricultural Statistics classifies as backyard farms—farms that hold no more than 20 head of adult-equivalent animals (BAS 2005). In the Agricultural Census of 2002, about 43 percent of all 4.8 million agricultural

[3]Based on 2000 estimates for disaggregated rural-urban poverty indices. Urban poverty incidence of families was 14.3 percent, whereas the incidence of national poverty for the same year was 27.5 percent. The latest estimate on poverty incidence of families in the Philippines for 2003 is 24.7 percent, with no disaggregation yet between rural and urban poverty.

Figure 1.1 Trends in share of backyard hog inventories, 1990–2003

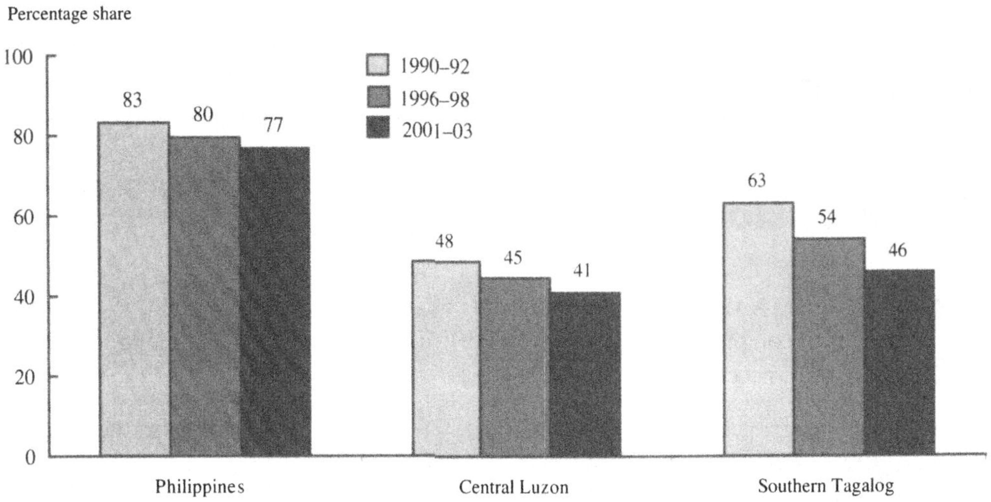

Source: BAS 2005.
Note: Shares are computed over three-year averages.

farm households in the country reported raising hogs, holding on average 4.2 head of pigs per farm (NSO 2002, 2005).

Despite the dominance of smallholders in pig production, growth in domestic demand for pork, however, is not associated with growth in smallholder market share, particularly in those markets experiencing the most rapid growth. Demand is concentrated in the major urban and commercial center—Metropolitan Manila—which accounts for 13 percent of the country's population; in 2003, the average income of households in this region was more than twice that of households outside the national capital (NSCB 2005). More peripheral regions, such as Central Luzon (Region 3; see Figure 1.2 for an explanation of the regions) to the north and the CALABARZON[4] area of Southern Tagalog (Region 4) to the south and east of the national capital, form secondary growth centers of consumer demand stemming from progressive urbanization and constitute a further 15 percent and 10 percent of the country's population, respectively. Moreover, average household incomes in these two regions were, respectively, about 40 percent and 20 percent higher than the average for households outside Metropolitan Manila (NSO 2005). Smallholder pig producers in these two regions face large and growing meat markets in the national capital and other major urban centers within these regions, yet the competition for market share with larger commercial pig producers is also especially intense, as these are the regions targeted by the big players.

The competition for market share in the domestic pig industry, indeed, appears to be different at the national level on the one hand, compared to the major pig-producing regions of Central Luzon and Southern Tagalog, as shown in Figure 1.1. Even as total hog inventories were experiencing significant growth (more rapidly so in Central Luzon and Southern Tagalog, both at about 4.5 percent per year), the share of backyard hog inventories was declining at a faster rate in these two regions than elsewhere. Moreover, although the backyard share at the national level continues to be high at 77 per-

[4]This is an acronym that refers to Cavite, Laguna, Batangas, Rizal, and Quezon, in that order: the five provinces of Southern Tagalog that make up Region 4a, excluding the island provinces to the south.

Figure 1.2 Regional distribution of the number of registered commercial pig farms, 1997 and 2004

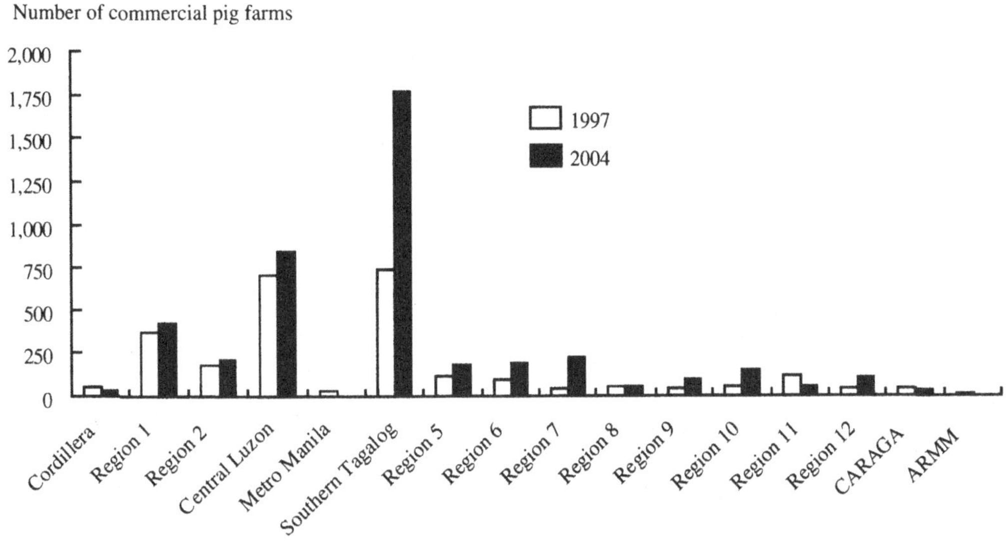

Source: BAS 2004.

Notes: A region is a subnational administrative unit in the Philippines made up of several provinces having the same geographical features and roughly homogeneous characteristics. The regions are defined as: Region 1, Ilocos; Region 2, Cagayan Valley; Region 3, Central Luzon; Region 4, Southern Tagalog; Region 5, Bicol; Region 6, Western Visayas; Region 7, Central Visayas; Region 8, Eastern Visayas; Region 9, Western Mindanao; Region 10, Northern Mindanao; Region 11, Southern Mindanao; Region 12, Central Mindanao. ARMM indicates the Autonomous Region of Muslim Mindanao; CARAGA comprises Agusan del Norte, Agusan del Sur, Surigao del Norte, and Surigao del Sur; Cordillera is the Cordillera Administrative Region.

cent, backyard producers in these two regions now account for less than half of production.

The trend in commercialization of pig farms, particularly in Central and Southern Luzon, can also be gleaned from the significant increase in the number of registered commercial hog farms, as estimated by Bureau of Agricultural Statistics in 1997 and 2004 and presented in Figure 1.2. The number of commercial hog farms increased by about 60 percent over a 7-year period. They are concentrated in the two regions close to Metropolitan Manila. In Southern Luzon, the number of commercial farms more than doubled. The average swine herd size at any one time on these commercial farms was 430 head in Southern Luzon and 1,100 head in Central Luzon (BAS 2004).

Although the number of commercial pig farms was increasing, particularly in Southern and Central Luzon, the total number of farms reported to be raising pigs decreased, including both part-time and full-time operations. From the Census of Agriculture and Fisheries of 1991 and 2002, the decline in the number of farms raising pigs is shown in Figure 1.3 for Southern and Central Luzon (NSO 2005). As the number of pig farms declined, the average number of pigs per farm increased, more than doubling from 7 to 16 head in Central Luzon, from 5 to 7 in Southern Tagalog as a whole, and from 8 to 10 head in the CALABARZON subregion (NSO 2005).

Thus even as the market for domestic pork as a whole has been expanding rapidly since 1990, more farms were exiting the activity than entering. Given the significant increase in the number of registered commercial farms in both Southern Tagalog and Central Luzon between 1997 and 2004, it is clear that the producers that were exiting were smallholders.

Figure 1.3 Changes in the number of farms raising pigs in Central Luzon and Southern Tagalog, 1991 and 2002

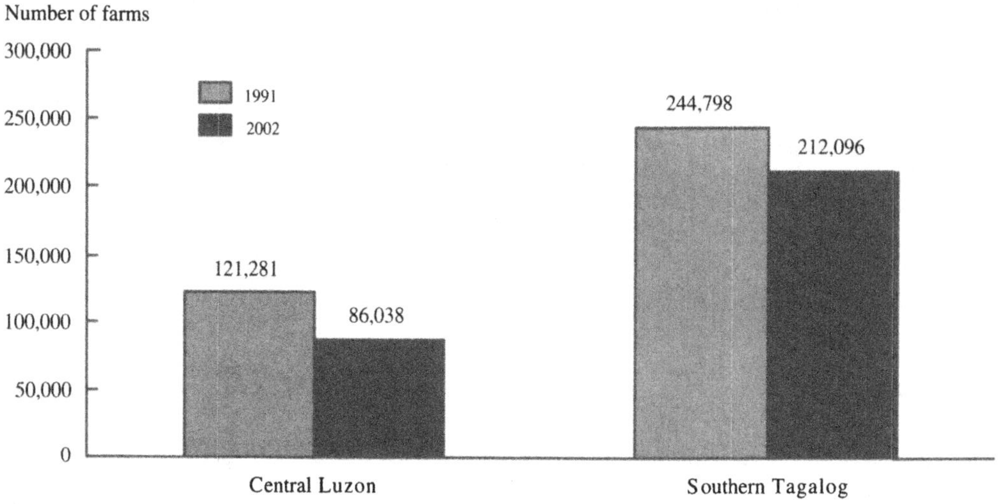

Sources: NSO 1991, 2002, 2005.

Changing Demand Patterns and Value Chains

The food consumption patterns of higher-income households of the national capital and surrounding urban centers are expected to be structurally different from those of lower-income households in the more distant provinces. Not only is there greater consumption of pork in these urban centers, but the demand for convenience and quality is increasing. As the demand for meat moves toward greater product differentiation associated with these traits, access to markets for such products may not be easy for smallholders. Their product must not only meet norms, but procurement agents and purchasers must believe that it does. This requires credible branding and certification or other forms of market trust and reputation that smallholders find it hard to achieve with the increasing volume and anonymity of supply chains.

The National Meat Inspection Service, the agency tasked with maintaining sanitary standards for meat for domestic and international trade, grades all slaughterhouses through which livestock from various producers are processed according to public hygiene standards for the slaughterhouse itself, thereby determining whether meat processed through the facility in question is restricted to local markets (Class A), can be sold in national markets (Class AA), or can be exported to the international market (Class AAA). Private firms with their own slaughtering and processing facilities may set even higher standards than the public ones to capture client patronage for high-end markets. Among large pig farms, the objective of producing finished hogs with higher product acceptability in more remunerative markets is demonstrated by their choice of breeding stock and animal feed and the measures taken to promote animal health. Among large companies carrying brand names, whose sights are on niches in the export markets, efforts are made to gain recognition of their slaughterhouses and processing plants by the International Organization for Standardization and the Hazard Analysis and Critical Control Point. With international certification for food quality and safety of products, these companies obtain better access than do smallholders to large supermarkets and other formal chains in the domestic market that supply better grades of pork cuts.

Importance to the Future of Smallholder Farming

Hog raising is one of the few activities that small agricultural operators in Luzon can engage in whose product is growing rapidly in demand. Poultry in Southern Luzon has migrated out of the smallholder sector altogether, fruits and vegetables are seasonal, and rice demand is flat.[5] Thus trends in the scaling-up of hog production and smallholder exit raise questions about the continued ability of smallholders to compete for this growing market. These concerns are amplified by at least anecdotal perceptions of changes in demand attributes, such as certified food safety, and in the concentration of vertically coordinated market chains supplying the growth segments of the market. Under these conditions, the inability to compete in smallholder hog production may precipitate the exit from smallholder agriculture altogether.

Although moving to partial or full-time nonfarm employment seems inevitable for many smallholders in Luzon and eventually elsewhere in the Philippines, the ability of urban areas and nonfarm activities to absorb large numbers of people is limited. In particular, growth in demand for nonfarm goods and services in rural areas depends on having widespread growth in the purchasing power of rural people, which comes from sales to places outside the local area (Hazell and Roell 1983).

Given its significance to understanding the future for smallholder farming in the Philippines, determining the causes of the scaling up of hog production in Luzon is crucial. In particular, it is important to determine whether the process is an exogenous one driven by the nature of technological change, or whether it is driven in part by organizational or other policy-relevant issues that government can influence. The objective is not necessarily to preserve smallholder agriculture but to stop aggravation of the problem caused by existing policy distortions favoring large-scale producers and to find market-oriented ways to prevent smallholder exit from being more rapid and disruptive than necessary.

Research Objectives

Can Smallholders Compete against Larger Farms?

This study cannot directly assess many important issues, such as the nature of changing demand for food safety, although they may be relevant for final outcomes for smallholders. However, it will look at how well a sample of smallholders in Southern Luzon can compete against larger specialized hog farms. Competitiveness as used here refers to the capacity to maintain or improve a market position. For producers, it is manifested by the ability to generate and sustain positive profits. A farm must produce at lower cost per unit of output than the price it receives for its output. In the long run, efficient firms—efficient in the sense of the profit generation—survive, driving the inefficient ones out of the market. Thus to survive, smallholders must be relatively efficient. However, they also need to live off a small volume of output to be able to generate sufficient profit to sustain themselves.

Relative unit profit efficiency is a useful yardstick for the purpose of this study. This quantity is defined as the combination of technical and allocative efficiency exhibited by specific farms with specific fixed resources and technology that face a specific set of input and output prices. It is expressed as a percentage distance of a specific farm with specific resources from the optimum that could be achieved by any farm with those resources and in that price and technology environment (Coelli, Rao, and Battese

[5]FAO Food Balance Sheets show that rice had grown only at an average of 1.4 percent per year from 1990 to 2002 (FAO 2005).

1998). As shown below, such a measure can be estimated from a sample of farms of different sizes, and it shows the relative unit profit efficiency of each farm relative to others in the sample. The observed scaling up of pig operations, particularly in Central Luzon and Southern Luzon, suggests a greater profit efficiency of larger-scale operations. This observation will be investigated empirically.

The study also looks at the unit profits of pig producers. If, on average, the unit profits of smallholders are higher than those of large farms, and if they are also more unit-profit efficient, then it is fair to say that smallholders can compete with large farms on a level playing field in production and marketing. Larger farms will find it hard to cut unit profits below smallholder levels and drive them out of the market under these conditions. If, however, large farms are more profit efficient, they will eventually cut smallholder profits to the point where smallholders will not be able to profit from pig farming, except perhaps as a hobby or secondary activity.

An empirical study can also shed light on why some farms are more unit-profit efficient than others. True economies of scale, such as savings from bulk transport of feed, may be one factor that explains the greater survival capacities of larger farms. However, farm-specific transaction costs that are greater for smallholders, hidden subsidies to larger farms, and scale-related negative environmental externalities from high-intensity pig production may work to increase the relative unit profitability of large farms in financial (not social) terms. Unlike true economies of scale, these drivers are fundamentally distortions that decrease social welfare at undistorted prices. As such, they would unduly hasten the departure of smallholders from pig farming and should be addressed by policy.

Thus the main objectives of this study are to understand the market forces, structural factors, and policies that affect the relative profit efficiency of small and large farms; to explain why some farms are more profit efficient than others; and use this information to see whether there is scope for remedying any bias of incentives against smallholders. Another objective involves understanding why some smallholders do not participate in hog raising at all, even if some smallholders can compete with large farms.

What Is the Role of Transaction Cost Barriers in Smallholder Participation and Performance?

Notwithstanding the apparent profitability of pig production as a sideline or a full-time occupation, some households will not engage in the activity because they may have sufficiently higher opportunity costs of time and resources to preclude engaging in the arduous and messy business. However, there may be households with low resource opportunity costs and a great need for income that are not participating, despite a desire to do so. They may be households that had engaged in pig farming before, but have been driven out of the activity after incurring sustained losses. They may also be households that have little resources of their own to start the activity, or have no access to external sources of financing. Understanding why some smallholders participate in market-oriented pig production and others do not is thus important to helping resource-constrained poor people in Luzon. In particular, it would be relevant to know whether transaction costs affect the efficiency of smaller producers more than they do those of larger farms.

Transaction costs in their purest sense are the costs of exchange that arise from asymmetries across market actors in access to information (Williamson 1989). Transaction costs[6] arise when buyers and sellers do

[6]Transaction costs include those of searching for a trading partner and for the "best" price, of screening potential trading partners to make sure they are reliable, of negotiating and bargaining with potential trading partners

not have the same information about what is being sold at the time of sale and what the alternatives are. For example, transaction costs arise when the buyer is not sure of the percentage of back-fat of a hog or the feed-dependent flavor of meat being sold until the animal is slaughtered, or when a buyer cannot be sure of the true quality of feed (knowing only the list of ingredients on a feed bag) being purchased.

Transaction costs arising from asymmetries in access to information are especially prevalent in the livestock product business and represent net social losses in the sense that neither buyers nor sellers gain from their presence (Williamson 1989). They are real costs that are incurred as lower prices received or higher prices paid than the ideal. Transaction costs may not be directly observable, but one clue to their presence is manifested when prices received by large farms differ from those received by small producers, even after considering differences in transport costs and in observable product quality. Thus buyers without reliable information on the quality of output from a backyard producer are only willing to pay a lower price per unit than otherwise would have been the case. In such instances, lower prices imply higher transaction costs. The challenge is to show the extent to which transaction cost differences explain these differences in prices.

What Can Be Done about Disproportionate Transaction Costs?

If transaction costs described in the previous section affect small producers more than they do larger ones, then collective action by smallholders might help reduce the transaction costs they face. Contract farming is one such institutional arrangement that organizes production decisions and marketing of output. This study looks at this question through a case study (presented in the Appendix) of an interesting institutional innovation in smallholder contract farming of hogs in Southern Luzon.

---

to reach an agreement, of transferring the product (which usually involves transportation, processing, packaging, and securing title, if necessary), of monitoring the agreement to see that its conditions are fulfilled, and of enforcing (or seeking damages in case of default) the exchange agreement (Staal, Delgado, and Nicholson 1997).

CHAPTER 2

Previous Work and Methodology

A variety of approaches have been developed in the empirical literature to address the drivers of market participation of smallholders in developing countries; this study focuses on two. The first is a simple, direct approach, which estimates a reduced-form model of market participation using samples of participating and nonparticipating farmers. Implications are then deduced from the sign and significance of the coefficients of explanatory variables, recognizing that these are not structural models. These explanatory variables are typically household characteristics, but they are often chosen to proxy differences across farms in access to information or assets (see below). The second is an indirect approach using models that estimate the determinants of relative profit efficiency across farms in a sample. Under the hypothesis that relatively less profit-efficient farms will drop out over time, the significance and relative magnitude of these determinants are indicators of which factors will allow some farms to grow and others to wither. Both approaches can be adapted to include issues of scale, provided that care is taken to exclude endogenous variables as explanatory variables.

Direct Approaches to Estimating the Determinants of Market Participation

In many developing countries, smallholders may find it difficult to participate in markets. They are faced with a range of barriers and limitations that reduce incentives for participation. These may be reflected in hidden costs that make access to markets and productive assets difficult, if not impossible. A key class of these barriers are transaction costs, that is, observable and nonobservable costs associated with exchange, such as searching, bargaining, monitoring, and enforcing agreements in markets (Coase 1960; Williamson 1989; Holloway et al. 2000).

Transaction costs thus typically originate from asymmetries in information between buyers and sellers about potential contracting parties and the price and quality of the resources in which they have property rights (for example, personal time, travel expenses, and communication costs). Where there is no asymmetric information between buyers and sellers, transaction costs may still ensue in the form of hidden costs of bargaining with potential trading partners and officials to reach an agreement, making contracts (formal or informal), monitoring the agreement to see that its conditions are fulfilled, and enforcing the exchange agreement (Williamson 1989; Staal, Delgado, and Nicholson 1997; Holloway et al. 2000; Heltberg and Tarp 2001).

Significantly, some households incur higher transaction costs than others, and the household attributes responsible for this are often correlated with scale (de Janvry, Fafchamps, and Sadoulet 1991). Small farmers may not be able to get credit at a reasonable price if potential lenders cannot ascertain the farmers' ability to repay. Some products incur more transaction costs than others. For instance, the inability to easily observe the quality and safety of perishables, such as milk, at the time of sale limits product price in long-distance anonymous markets unless the farmers are part of a cooperative that, in effect, brands their product (Staal, Delgado, and Nicholson 1997).

Previous studies have investigated the issue of transaction costs in participation decisions of individuals or households. A significant body of early literature addresses rural finance (for example, Saito and Villanueva 1981; Cuevas and Graham 1986; Cuevas 1988; Zander 1992; Fenwick 1998) and labor market participation (for example, Evenson and Roumasset 1986; Lanzano and Evenson 1997). Relevant studies in agriculture specifically dealt with transaction costs and participation in output markets and milk markets. There is consensus in the literature that the very existence of transaction costs tends to inhibit market participation.

Goetz (1992) proposed that failure to participate in specific commodity markets results from high fixed transaction costs. Key, Sadoulet, and de Janvry (2000) subsequently showed that both fixed and variable transaction costs affect market participation decisions, but that only variable transaction costs are important in the household's decision of how much to supply to the market. Heltberg and Tarp (2001), however, used exogenous variables that are theoretically expected to determine the size of transaction costs, such as distance and type of transport, to determine variable transaction costs and information variables to determine fixed transaction costs. Their findings highlighted the importance of nonprice factors, such as technology, transport infrastructure, farm endowments, and area characteristics. It was also shown that policies that support the expansion of the number of market participants are far more important than those for stimulating farmers who are already in the market to increase their supply.

Livestock products produced on small farms under tropical conditions are especially subject to transaction costs. Lapar, Holloway, and Ehui (2003) examined the competing effects of transaction costs, labor mobility, capital (intellectual, financial, and physical) formation, and indebtedness on market participation and selling decisions by smallholder livestock producers. In contrast, Holloway et al. (2000) explored the impact of household-level transaction costs and the choice of production technique on the decision of farmers to sell milk to marketing cooperatives in the Ethiopian highlands. The issue of transaction costs associated with market distance is particularly important in milk markets, as shown in Staal, Delgado, and Nicholson (1997); Holloway et al. (2000); and Staal et al. (2000), and highlights the need for institutional innovations to address informational and physical market infrastructure deficiencies.

For this approach to be valid, alternate explanations of why some farmers participate —and others do not—must be factored in. Some households may not participate in hog raising simply because they have household-specific options in other economic areas that are more attractive: such possibilities need to be included in the analysis to avoid biasing the estimation. In part, these options can be controlled for by choosing samples of participants and nonparticipants that are very similar except for the fact of participation, on the assumption that their factor opportunity costs are similar. In part, option control requires ensuring that the explanatory variables adequately cover the items that drive participation, as in any regression. The use of reduced forms makes adequate coverage easier—because the included variables are correlated with many factors—but interpretation harder. Practically speaking, reduced-form participation equations are useful, but a better approach is needed—one that uses a structural model.

Models for Investigating Market Participation

Market participation can be modeled as a decision with an observable discrete outcome, such as a "yes or no" decision. This type of model is one of a category generally known as qualitative response models, where conventional regression methods are not appropriate; instead, maximum likelihood is the method of estimation (Maddala 1983; Greene 1997). The simplest of these models is that in which the dependent variable y is binary; that is, it can assume only two values, which for convenience and without loss of generality we denote by 0 and 1. For instance, y can be defined as 1 if the individual participates in the market and 0 otherwise. This model can be implemented statistically using a probit model, which assumes a normal distribution of its error terms, or a logit model, which assumes that the error terms have a logistic distribution. The debate as to which is the more appropriate model has remained unresolved on theoretical grounds; however, it seems not to make much difference in most applications (Greene 1997, p. 876).

Probit analysis of market participation is structurally based on the random utility model, in which a household or individual faces two levels of utilities, U_1 or U_0, from making choices between participation and nonparticipation, respectively. However, the observed state only reveals which choice provides a higher utility but not the unobservable utility. That is, the observed (latent) indicator equals 1 if $U_1 > U_0$ and 0 if $U_1 < U_0$. By assuming that differences across utilities are determined by household- or individual-specific characteristics, the model can be estimated by regressing an indicator variable —for example, $Y = 1$ for participants and $Y = 0$ for nonparticipants—on X, which is a matrix of household- or individual-specific covariates to obtain β. The vector β is composed of unknown coefficients controlling the relationship between household- or individual-specific characteristics and market participation plus a random error.

Transaction cost is one, if not the, key feature of recent works to empirically model market participation decisions. In a household-model framework (Singh, Squire, and Strauss 1986), market participation ensues when the shadow price (or opportunity cost) is equal to the market price, assuming zero transaction costs. When transaction costs are nonzero, as empirical studies have validated, then market participation is likely subject to a threshold below which households or individuals remain in autarky (that is, are nonparticipants).

Two-Step Selectivity Procedure for Estimating Market Participation

Ideally, the ordinary least squares model is applicable to determining factors that affect the level of participation in the market when all households participate. However, some households may prefer not to participate in a particular market in favor of another, whereas others may be excluded because of market conditions or household resource constraints. If the ordinary least squares regression is estimated while excluding the nonparticipants from the analysis, a sample selectivity bias is introduced into the model. Such a problem is overcome by following a two-stage procedure, as suggested by Heckman (1979).

In this study, therefore, the Heckman's two-stage selectivity model is used in investigating the factors that influence the probability of being engaged in hog production while simultaneously estimating the factors affecting the level of participation.

The selection model in this study is defined as:

$$QTYKILO_i = \beta_0 + \beta_1 X_1 + \beta_2 X_2 + \ldots + \beta_n X_n + e_1. \qquad (2.1)$$

QTYKILO can be observed if

$$\alpha_0 + \alpha_1 x_1 + \ldots + \alpha_m x_m + e_2 > 0, \qquad (2.2)$$

where QTYKILO$_i$, the quantity of output sold plus the unsold stock from farm i (in kilograms liveweight), is the dependent variable defined in equation (2.1). The determinants (X_i) are price of output, prices of inputs, and some transaction costs proxied by variables assumed to affect the level of output. These variables are discussed in detail below. The quantity β_i represents the estimated marginal effects of X_i, e_1 is an error term normally distributed with mean 0 and variance σ, and e_2 is an error term normally distributed with mean 0 and unit variance. The correlation of e_1 and e_2 is equal to ρ. When ρ is nonzero, standard regression techniques applied to equation (2.1) yield biased results. Equation (2.2) is the selection equation.

The first step of the Heckman procedure establishes the probability of participation in the output market. For the individual farmer, the decision to participate or not to participate in pig production can be formulated as a binary choice model that can be analyzed using the probit equation below. The empirical specification of the probit model to be estimated by maximum likelihood estimation is defined as:

$$PART_i^* = \alpha_0 + \sum_{j=1}^{m} \alpha_j x_j + e_2 \qquad (2.3)$$

where each x_j is an exogenous variable (each defined below), which is assumed to influence the participation decision; e_2 is as defined above; the α_j's represent estimated marginal effects of the determinants of participation; PART is the dummy variable that takes the value of 0 or 1, determined by PART* as follows: PART equals 1 if PART* >0; otherwise, it equals 0. The probit functional form compels e_2 to be homoskedastic because the form of the probability depends only on the difference between the error term associated with one particular choice and another (Amemiya 1981).

The marginal effects are estimated on variable means. This calculation involves taking the partial derivatives that measure the change in the probability of participation per unit change in the independent variable.

Variables Used in the Model

According to the literature, transaction costs, among others, are important factors that reduce the incentives for households to participate in markets for various agricultural commodities. These transaction costs, however, are often not directly observable but can be represented by household characteristics indicating capacity to deal with particular transaction cost barriers to market entry.

In applying the Heckman two-stage selectivity model to the sample of pig-producing smallholders and nonpig-producing households in the study, the structure of the resulting sample is taken into account. An important feature of the sample is that none of the households engaged in pig raising were producing solely for home consumption—all were market participants. Thus, for the entire sample, the observable decision to produce pigs cannot be separated from the observable decision to engage in the market for pigs. In the application and specification of the model, therefore, engagement in pig production implies market participation; that is, the decision to produce is simultaneous with the decision to sell in the market.

The specification of the model is guided by the general hypothesis that household resources, opportunity costs of the members, market prices for outputs and inputs, and other household characteristics linked to individual household access to assets, resources, and information (which enhance their capacities to deal with transaction costs of market entry) influence both the decision to participate in pig production and the level of market participation. According to the literature on market participation, there are two types of transaction costs—fixed and variable costs. Fixed transaction costs refer to those incurred once the decision to participate is made, and their levels remain invariant to the size of the operations. They are often related to the costs of transport, market search, negotiation and bargaining, screening, enforcement, and supervision (Key, Sadoulet, and de Janvry 2000). Variable transaction costs, however, vary with the level of market transactions. Fixed transaction costs may affect only the decision to participate, whereas variable transaction costs may affect both the decision to participate and the level of participation (for examples, see Key, Sadoulet, and de Janvry 2000; Bellemare and Barrett 2006). Specifically, the effects of fixed and variable transaction costs on the decision to participate and the extent of participation depend on assumptions about the nature of household decisionmaking; that is, whether households make the choice to participate and determine the level of participation simultaneously or sequentially, as illustrated by Bellemare and Barrett (2006). For example, Bellemare and Barrett (2006) have shown that when the choice to participate and the level of participation are sequential, only the variable transactions costs appear to affect the decision of how much to sell in the market.

The circumstances under which market participation is analyzed in the present study suggest that market participation decisions are sequential; that is, households decide whether to participate in hog production and then determine the level of output to produce for sale in the market. Hence, it is hypothesized that the decision to participate may be determined by a different set of variables than those determining the level of market participation, where the latter may be less affected by fixed transaction costs than by variable transaction costs and market prices. It should also be noted that the data set in this study includes pig producers and nonpig producers, where information on the former, such as production levels and prices for inputs and outputs, are not available for the latter set of respondents. Although prices for output and inputs are reported by participants over the production period, these prices cannot be presumed to be the relevant prices that nonparticipants would have faced in earlier investment periods that would have influenced their decisions to participate. Hence, the absence of appropriate prices for nonrespondents limits the set of covariates that are common to both types of respondents that can be included to estimate the participation models. Nonetheless, the theoretical underpinnings of the choice of variables in the decision to participate and the level of participation are given due attention, as in de Janvry, Fafchamps, and Sadoulet (1991); Key, Sadoulet, and de Janvry (2000); and recently in Bellemare and Barrett (2006).

The structure of the model for estimation purposes, therefore, consists of four sets of variables. The first set refers to the household resources available for pig production. The second set pertains to the opportunity costs of the household in the use of household resources for engagement in this economic activity. These first two sets are hypothesized to influence both the decision to participate and the level of market participation.

The third set of variables pertains to the opportunities presented by the market in terms of the prices of output and major inputs in pig production. Theoretically, these prices influence both the decision to participate and the level of market participation. They affect the decision to participate in the way that potential entrants to the economic activity evaluate the expected net gains prior to undertaking the investment in the activity. Their effect on the level of participation follows from the standard impact of input and output prices on supply behavior. In the model specification, however, the absence of household-level data on suitable prices faced by nonparticipating households precludes the inclusion of prices in the estimation of the decision to participate, so that prices appear only in the second stage of the model estimation. In the context of the prevailing market conditions for hogs in Southern Luzon, the exclusion of prices in the decision to participate in or opt out of hog production may be justified because producers are essentially price takers in the market for hogs. Thus, the preconditions for market entry that determine potential entrants' decisions may be their capacity to engage in the activity (for example, human, physical, and financial household resources) and to meet the various start-up costs required in terms of time and effort.

The fourth set of variables refers to factors that represent the household's ability to deal with transaction costs related to smallholder market-oriented pig production. This set of variables is further categorized into those factors that affect fixed transaction costs, which are likely to influence the decision to participate, and those factors that affect variable transaction costs, which are likely to affect both the decision to participate and the level of market engagement.

The specific variables in the model are discussed according to their categorization in the four sets of variables described above. In the first set of variables, family labor of working age is the main household resource, particularly in smallholder livestock production. This resource is represented by the variable household size (HHSIZE), the number of household members more than 15 years of age. High values of HHSIZE are expected to contribute positively to the likelihood of deciding to participate, as well as the level of participation. A large number of working-age household members represents an advantageous resource for a labor-intensive activity such as pig production.

The second set of variables includes the employment status of both the household head and the spouse. Although the study asserts that smallholder pig production presents employment and income opportunities for households, there are also opportunity costs to engaging in such a labor-intensive (and sometimes messy) activity at the household level. Given the important role of family labor in a household-based enterprise, a critical opportunity cost is the nature of employment of the household head as well as that of the spouse. The possibility of this cost is captured by the employment of the household head or the spouse in the formal sector of the economy that may constrain the time available to them to engage in pig production. These are represented by dummy variables for the main occupations of the household head and spouse (HHGOVPVT and SGOVPVT, respectively); a value of 1 for either of these variables signifies employment in government or private enterprise. The hypothesis is that high opportunity costs faced by the household head or spouse in terms of available time will negatively influence the decision to participate in market-oriented pig production and the level of that participation.

The third set of variables includes the market prices of output and inputs in pig production that are hypothesized to capture the effects of market forces that influence ecomonic incentives of market participation. Specifically, this set of variables includes price of output of slaughter hogs (pesos per kilogram liveweight; PSLHOG), price of weaners used as inputs to hog production (pesos per kilogram liveweight; PWEAN), and price of feed (pesos per kilogram; PFEED). It is hypothesized that the output price influences participation positively, whereas higher input prices have negative effects on participation.

The fourth set of variables in the model contains the factors influencing the capacity of the household to deal with transaction costs related to market-oriented pig production and are specifically associated with fixed and variable transaction costs. The variables influencing household-specific fixed transaction costs in market entry are education level of household head (years of formal schooling; HHEDUC); gender of the household head (1 = male; HHGENDER), represented by a dummy variable; existence of agricultural land owned or cultivated (1 = yes; OWNALAND); existence of a secondary income source for the household head other than hog raising (1 = yes; HHOTHR), represented by a dummy variable; existence of remittances from overseas (1 = yes; VREMITY), represented by a dummy variable; and a dummy variable for connection to a piped-in water source (1 = yes; WATER).

A higher level of education of the household head represents increased capacity to process production technology-related and market-related information and better negotiation skills in market transactions. Thus, the higher the level of education of the household head, the more likely the household would participate in the economic activity.

In the social-cultural context where males are dominant in entrepreneurial transactions, being a male signifies a stronger position in negotiations and bargaining. Thus, it is hypothesized that male-headed households would more likely participate in market-oriented pig production.

The ownership of an asset, land, that is generally useful as collateral bestows an advantage to the household in accessing credit, when there are barriers to sources of external financing. Thus, households that own or operate an agricultural land would be better able to participate in a pig-raising enterprise. Having land available for pig pens (and manure spreading) is also a requirement for engaging in pig production, which will likely bar landless households from the activity.

The household head's having a source of income from other occupations represents access to additional funds for engaging in pig production. Where there are barriers to access to capital and credit in the financial markets, the availability of an additional source of income from remittances could provide leverage for engaging in a market-oriented activity, such as pig raising.

The access to piped-in water is hypothesized to facilitate operations in pig production because pigs require a constant supply of fresh water for drinking and for bathing to cool down in tropical conditions. In addition, pig pens need to be cleaned regularly to sanitize the living conditions and keep the pigs healthy. The access to this service would thus encourage participation.

Finally, among those factors that are hypothesized to influence the household-specific variable transaction costs affecting market participation, the household's level of investment in breeding stock and the access to output markets are considered to be important. The level of animal stock, standardized as the value of the breeding stock (VSTOCK), captures transaction costs that may be scale-related. It is thus hypothesized that the larger the scale of investments in stock, the smaller will be the transactions cost per additional unit of output and thus the higher the level of market participation. The extent of market reach of the smallholder is represented by a dummy variable for market outlet (OUTLET), which has a value of 1 when market outlet is limited within the village. Under such circumstances, the buyers of output are neighbors, village traders, or village agents. Because no information on breeding stock and market outlets is available for nonparticipants in the sample, these variables could not be included in the decision equation (the first stage of the Heckman model), although theoretically, they may likely influence both the decision to participate and the level of market engagement. Thus, in the specification of the model, these variables are only found in the second stage of the model.

Indirect Approaches to Estimating Incentives for Increased Market Participation

Issues in Economic Efficiency

Economic efficiency analysis tries to measure such factors as the efficiency of a farm or household relative to some standard, the performance of each farm or household relative to other farms or households in the sample units, the sources of differences in relative performance, and the deviation of each farm or household from the optimal level. It also allows for identification of the drivers of efficiency in the sample farms or households.

Analyses of technical and allocative efficiencies are specific methods within this group of analytical approaches to evaluate economic efficiency. Technical efficiency relates to obtaining the maximum quantity of output for given amounts of physical inputs. Allocative efficiency involves minimizing the cost of inputs for a given amount of output at specified input prices. Profit efficiency is an amalgam of the two, conveyed by the maximization of farm profits given a set of input and output prices, including shadow prices for fixed farm resources (Fried, Lovell, and Schmidt 1993; Coelli, Rao and Battese 1998).

Methodology for Investigating Farm-Specific Profit Efficiency

A standard way of assessing farm-specific relative profit efficiency is to estimate a "profit frontier" across a sample of farms and then to measure how far each farm in the sample lies below the frontier (Ali and Flinn 1989; Coelli, Rao, and Battese 1998). Conceptually, such a frontier can be thought of as a function mapping profit per unit to relative input and output prices and quantities of nontraded factors of production, where each point is the maximum profit per unit that a farm can achieve given those relative prices and access to resources. Given a set of prices, the average farm with that level of resources will fall on or below the frontier. Thus, an ordinary least squares regression on data from a sample of farms of different sizes of profits per unit of output against input and output prices and fixed factors of production (for example, land and labor) will always lie below the theoretical frontier. The frontier itself has to be estimated in some fashion by looking at data for farms that perform best at each level of resources. Several approaches to estimating this efficient frontier are described by Fried, Lovell, and Schmidt (1993).

The measurement of the most efficient state can be improved by estimating a stochastic profit frontier, which allows for measurement error in the econometric estimation of the frontier itself and thus accounts for those farms observed to lie above the estimated "best" frontier (see Battese 1992 for a survey of this literature). In the present case, the dependent variable is profit per unit of output, and the explanatory variables are farm-specific fixed resources (land, family labor, sunk capital), farm-specific input prices (feed, medicines, stock, and the like), and farm-specific output prices. In the developing-country situations studied, farm resources, such as land, may be nontradable inputs and must be accounted for in the frontier in terms of the amount available and not in terms of price. The unit prices received for output and prices paid for inputs can also be expected to vary greatly and reflect (and control for) quality differences and differential transactions costs, such as bargaining power and risk.

The actual performance of each farm in terms of unit profit can be compared to an ideal unit profit for that farm, given its resources and the prevailing input and output prices. The difference between the ideal and the actual profit per unit for that farm is the farm's relative profit inefficiency. Following Ali and Flinn (1989), Figure 2.1 traces a profit frontier for a sample of farms. Each point in the figure corresponds to the actual outcome in terms of profit per unit for a specific farm; points on the stochastic frontier curve (estimated by maximum likelihood methods) are optimally efficient farms (on the frontier), and all points below are inefficient farms in terms of their specific resources at prevailing input prices.

Farm-specific profit efficiency (deviations below the frontier) are measured as the ratio of actual profit per unit (Y_i in Figure 2.1 for farm i) and ideal profit (Y^*). Note that the curve denoting average profit for any given level of resources (shown as the locus of points Y in Figure 2.1)—estimated by ordinary least squares regression—is less than the ideal profit. The measure of farm efficiency embodied in Y_i/Y^* is bounded by 1 (best; on the frontier) and 0 (worst; no profit). Farm-specific inefficiency is the distance below the frontier, $Y^* - Y_i$.

If small farms have on average significantly higher profit efficiency per unit of output than large farms when family labor

Figure 2.1 Frontier stochastic profit function for a sample of farms

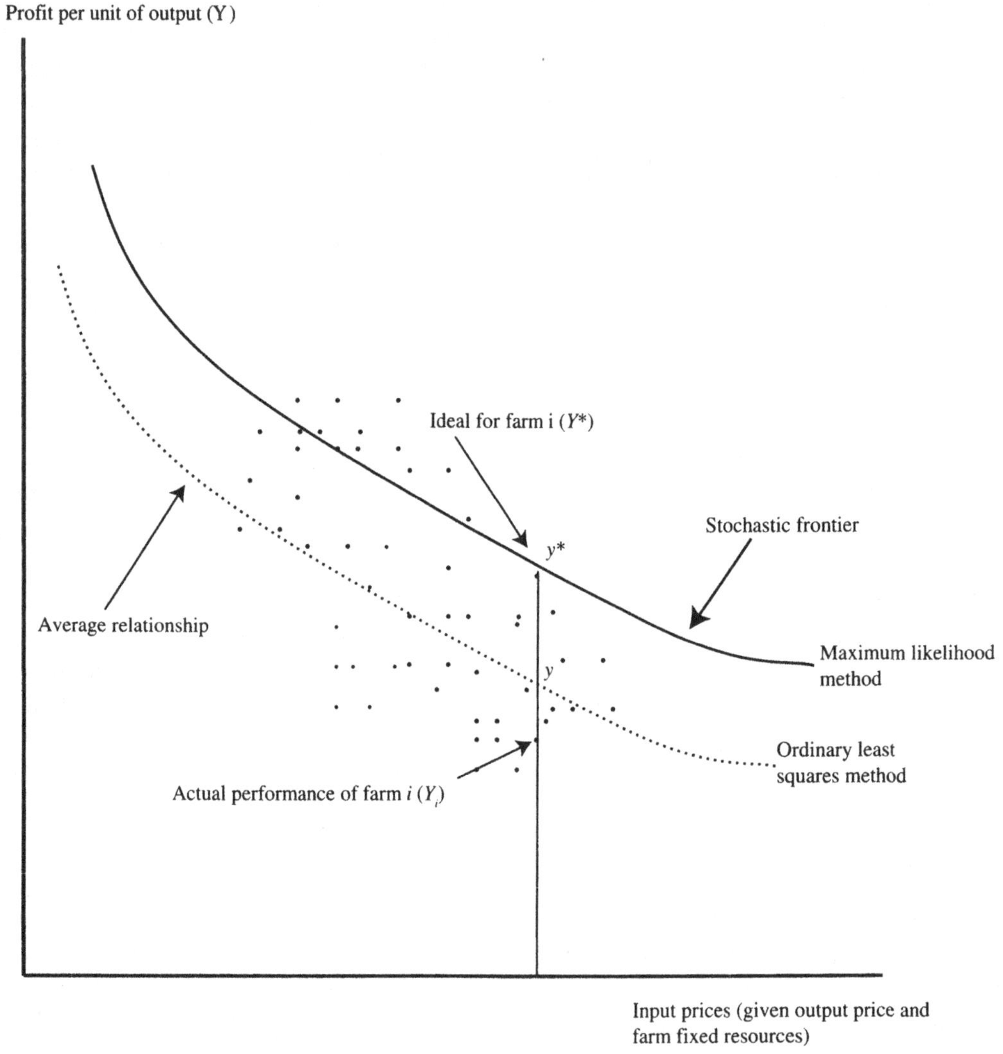

Source: Ali and Flinn 1989.

is not costed (under the assumption of low opportunity cost), then there is hope for survival of the smallholder. If smallholders are more efficient users of farm resources than are large farmers even when family labor is costed, then the outlook for smallholders is quite good. However, the methodology allows going beyond simply making this determination; it also permits the investigation of which elements contribute most to explaining relative unit profit efficiency for large and small farms. Individual farms, large or small, may lie well below the profit frontier for reasons other than technical or allocative inefficiency: farm-specific transactions costs or policy distortions may also influence their position relative to the frontier.

Working with a Stochastic Profit Frontier to Assess Drivers of Scaling-Up

The discussion above leads us to the principal methodological approach, which is to estimate a stochastic profit frontier and derive a farm-specific measure of relative inefficiency in securing profit per unit of output. These farm-specific measures of profit inefficiency can be taken as indicators of farm-specific competitiveness over the long run.

Precedents

Following the lead of Jondrow et al. (1982), Ali and Flinn (1989), and Battese and Coelli (1995), the relative profit inefficiency of each farm is explained in terms of the transaction cost barriers and farm-specific policy distortions faced by that farm. The first stage of the approach thus explains each farm's unit profit performance in terms of technical and allocative efficiency; the second stage explains differences in efficiency in terms of farm-specific differences in transactions costs and policy distortions. A review of empirical applications of frontier production models in agricultural economics is presented in Battese (1992).

Kalirajan and Flinn (1983) were the first to outline the methodology by which the individual farm effects can be predicted (as discussed by Jondrow et al. 1982) and applied the approach in their analysis of data on 79 rice farmers in the Philippines. A translog stochastic frontier production function whose parameters were estimated using maximum likelihood was assumed to explain the variation in rice output in terms of several input variables. The Cobb-Douglas model was found to be an inadequate representation for the farm-level data, and so a translog stochastic frontier production function was estimated to explain variations in rice output in terms of several inputs. The individual technical efficiencies ranged from 0.38 to 0.91. The predicted technical efficiencies were then regressed on several farm-level variables and farmer-specific characteristics. Similar approaches in subsequent work were taken by Huang and Bagi (1984), Kalirajan (1984, 1989, 1991), Kalirajan and Shand (1986), and Taylor and Shonkwiler (1986).

To ascertain differences across scale, Huang, Tang, and Bagi (1986) adopted a stochastic profit function approach to investigate the economic efficiency of small and large farms in two states in India. The variability of farm effects was highly significant, and individual farm economic efficiencies tended to be greater for large farms than for small farms (the average economic efficiencies being 0.84 and 0.80 for large and small farms, respectively).

Ngwenya, Battese, and Fleming (1997) used farm-level data for the 1988/89 agricultural year for a sample survey of wheat farmers in Eastern Free State, South Africa, to estimate stochastic frontier production functions in which technical inefficiency effects are modeled in terms of the size of the farming operation and other explanatory variables. The technical inefficiency effects were found to be negatively and significantly related to the size of the farms.

Huang, Tang, and Bagi (1986) and Ali and Flinn (1989) were the first to apply stochastic frontier estimation to estimate a profit frontier, unlike previous applications that estimated production frontiers. Ali and Flinn (1989) specifically estimated a stochastic profit frontier of modified translog type for Basmati rice farmers in Pakistan's Punjab. After estimating the technical efficiency of individual farmers, the losses in profit from technical inefficiency were obtained and regressed on various farmer- and farm-specific variables.

Kumbhakar, Biswas, and Bailey (1989) used a system approach to estimate technical, allocative, and scale inefficiencies for Utah dairy farmers. The stochastic frontier production function, which was specified, included both endogenous and exogenous variables. The endogenous variables included labor (both family and hired labor) and capital (the opportunity cost of capital expenses on the farm), whereas the exogenous variables included level of formal education, off-farm income, and measures of farm size for the farmers involved. Both types of explanatory variables were found to have significant effects on the variation of farm production. Technical efficiency of farms was found to be positively related to farm size.

The empirical model that is deemed most relevant and amenable to the investigation of the issues of the present study is the one by Ali and Flinn (1989), expanded by Battese and Coelli (1995). The approach here is in two stages: The first stage explains each farm's unit profit performance in terms of technical and allocative efficiency, and the second stage explains differences in efficiency in terms of farm-specific differences in transactions costs and policy distortions. The stages are described in the following section.

Defining a Measurable Index of Relative Competitiveness

A necessary condition for smallholders to be relatively competitive under developing-country conditions is the ability to produce at a lower unit cost of production than large-scale farms, with smallholders' family labor not being costed.[7] From the perspective of profit generation, the analogous condition is the ability of smallholders to realize higher profit per unit of output than large-scale farms, when smallholders' family labor is also not costed. Large-scale farms have the ability to remain in business with very thin profit margins because they can make up for low margins through large sales volume. For smallholders, very low per unit profits coupled with a small sales volume may not provide enough income to stay in business, and as large-scale farms expand their production, market prices will fall, thus squeezing smallholders out of the market. Thus, on the cost side, if large farms have lower per unit costs of production than do smallholders, even if the latter do not cost their family labor, then there is little hope for smallholders in this activity, except perhaps as a hobby or for minor income supplementation.

A sufficient condition for small farmers to stay in business—and perhaps to gain market share—is that they be more efficient technologically (by being on the production possibility frontier for existing technology) and allocatively (by being at the optimal point of factor combinations on the production frontier, given prevailing prices) in the use of farm resources. If small farms are more efficient users of farm resources in securing profits, then they have a cost advantage over large-scale producers that will be difficult to displace. This concept yields a measurable index of relative competitiveness: relative farm efficiency in securing profit per unit of output. Other things equal, farmers that are more efficient users of farm resources to secure profits per unit of output are more likely to be able to maintain market share than are larger producers who are less efficient in the same sense. Over time, the more efficient farmers are in a position to invest more into the farm enterprise and to grow, whatever their starting size.

Estimation of the Stochastic Profit Frontier Model

First Stage: The Stochastic Profit Frontier

The first stage involves estimating a stochastic profit frontier in the usual way, thus deriving a farm-specific measure of relative inefficiency in securing profit per unit of output. The stochastic profit frontier used in this study adopts the flexible transcendental logarithmic specification defined as:

$$\ln Y_i = \ln \beta_0 + \underset{\text{(dummies)}}{\sum_{t=1}^{T} \beta_{it} D_{it}} \tag{2.4}$$

$$+ \underset{\text{(input/output prices)}}{\sum_{j=1}^{J} \beta_{ij} \ln W_{ij}} + \underset{\text{(factors)}}{\sum_{k=1}^{K} \beta_{ik} \ln Z_{ik}}$$

$$+ \underset{\text{(price-factor interactions)}}{\frac{1}{2} \sum_{k=1}^{K} \sum_{j=1}^{J} \alpha_{ikj} \ln Z_{ik} \ln W_{ij}}$$

$$+ \underset{\text{(factor interactions)}}{\frac{1}{2} \sum_{k=1}^{K} \varphi_{ik} \ln Z_{ik} \ln Z_{ik}}$$

[7]This condition would be sufficient if smallholders' labor and other nontraded factors were fully costed.

$$+\frac{1}{2}\sum_{j=1}^{J}\theta_{ij}\ln W_{ij}\ln W_{ij}$$

(price-price interactions)

$$+ v_i - u_i,$$

(random error)
(random profit inefficiency effect)

where $\ln Y_i$ is the natural logarithm of the normalized profit per unit of output of the *i*th farm, defined as

$$Y_i = (\text{TR}_i - \text{TVC}_i)/\text{QTYKILO}_i, \quad (2.5)$$

where Y_i is the net annualized profit per unit of output for farm *i*; TR_i is the annualized total revenue (including value of unsold stock) from pig production per farm *i*; TVC_i is the annualized total variable costs per farm *i*, including depreciation of securing TR_i, excluding family labor; and QTYKILO_i is quantity of output per farm *i* as previously defined.

The quantity $\ln W_{ij}$ is the log of price of output and of input *j* (W_{i1}, denoted as PSLHOG, is the price of outputs [slaughter hogs or piglets]; W_{i2}, denoted as PFEED, is the price of feeds; and W_{i3}, denoted as PWEAN, is the price of weaners) used by the *i*th farm; $\ln Z_{ik}$ is the logarithm of household fixed factor *k* to control for differences in farm resources used by the *i*th farm (Z_{i1} is the value of breeding stock per unit of output denoted by VBRQ, Z_{i2} is the value of buildings and equipment per unit of output denoted by VBEQ, and Z_{i3} is total farm labor in hours per unit of output denoted by FLABHRS). The D_{it} are dummy variables that include: COMBINE-dummy to allow for fixed differences in mean unit profits among activities on the same farm with a value of 1 if the activity is a combination of farrow-to-wean, farrow-to-finish, and/or grow-to-finish where both piglets from own sows and weaners purchased are grown and sold as finished slaughter hogs; SORO-dummy for type of grower with a value of 1 if a household is under contract with Sorosoro Ibaba Development Cooperative (SIDC). The v_i, u_i are error terms, random error (v) is distributed two-sided, and inefficiency (u) is one-sided (downward) relative to the frontier. The quantities α_{kj}, φ_{kk}, β_{jk}, and θ_{ij} are coefficients to be estimated by the method of maximum likelihood using Frontier 4.1 (Coelli 1992).

Normalizing by output quantity builds in an assumption of constant returns to scale, so to allow for the fact that larger producers may in fact be using higher-grade technology or management than smaller farms, we need to control for Hicks non-neutral technical change on the right hand side of the first stage of the stochastic profit function. To deal with this problem, the feed conversion ratio is included to allow for the parameterization of Hicks non-neutral technological change, which is probably closely correlated with both technology and managerial ability. The weighted feed conversion ratio per farm is calculated by dividing total feed used (in kilograms) with total output (in kilograms of liveweight).

Second Stage: Inefficiency Determinant Model

The profit inefficiency effects (u_i)—assumed to be independently but not identically distributed—generated in equation (2.4) are estimated as:

$$u_i = \delta_0 + \delta_1 X_1 + \delta_2 X_2 \ldots + e_i, \quad (2.6)$$

where X_i is the *i*th farm characteristic determining relative inefficiency. The inefficiency error term e_i is defined by the truncation of the normal distribution with mean equal to 0 and variance σ^2. The truncation of e_i occurs at $+e_i \geq X_{ik}\delta_k$ (Battese and Coelli 1995). Equation (2.6) is simultaneously estimated with equation (2.4) using Frontier 4.1 software (Coelli 1992).

Following Ali and Flinn (1989), this study hypothesizes that characteristics of pig-producing households such as those that are linked to access to assets, resources, information, and markets are instrumental in explaining differences in profit efficiency among smallholders. Specifically, the variables are defined as: farm has access to feed of known quality (1 = yes; GOODFEED), a dummy variable; number of veterinarian visits to the farm per year at the time of the survey (VETVISIT); farm has piped-in water (1 = yes; WATER); mixed farmer occupation (1 = farmer also operates crop land; AGLAND), a dummy variable; existence of secondary income source for household head other than the hog raising business (1 = yes; HHOTHR), a dummy variable; existence of major income for spouse of household head other than the hog raising business (1 = yes; SPOCC), a dummy variable; household receives remittances from abroad (1 = yes; VREMITY), a dummy variable; and a dummy for market outlet being a neighbor, village trader, or agent (1 = yes; OUTLET).

The hypothesized effects of the specified household and farm characteristics on relative efficiency of the pig farms are discussed below. The variable representing having reliable access to (and confidence in) feed of known quality (GOODFEED) tends to improve farm efficiency because the farmer spends less time securing quality feed, thus lowering the cost of information search. Access to veterinary health services (VETVISIT) leads to lower livestock mortalities, more healthy pigs for the market, and buyer recognition and confidence that output supplied by the farmer is disease free. Thus, access to veterinary services is hypothesized to improve farm efficiency. The access to infrastructure and services that lead to reliable water supply (WATER) is vital to market-oriented pig production for the reasons given earlier in this chapter. The variable WATER is thus hypothesized to contribute positively to farm efficiency.

The following group of household characteristics are asserted to represent access to assets relevant to pig production and to resources that improve financial liquidity, which can be drawn upon by the household when needed for continuous operations. Being a mixed farmer operating on an agricultural land (AGLAND) implies the availability of a physical area directly usable for operation or expansion of pig production. It also implies the availability of land to facilitate the disposal of waste (for instance, manure as fertilizer for crops) generated in pig production, or the use of income generated from crops to bridge the differences in the income and expenditure streams in pig production. The existence of a secondary source of income of the household head (HHOTHR), or of the spouse (SPOCC), and the existence of remittance income from abroad (VREMITY) also improve the household's overall liquidity position throughout the year. Thus, access to these types of assets and resources are hypothesized to improve the efficiency of the farm, all other things being equal.

Finally, access to reliable output markets significantly affects the efficiency of smallholder producers. Market outlets are categorized as either within the village (OUTLET) or outside it. It is hypothesized that outlets limited to the village are a disadvantage to smallholders who sell to them. Thus, marketing to buyers within the village tends to limit market outlet choices and reduce farmer efficiency in generating income and profits.

CHAPTER 3

Participation and Nonparticipation in Market-Oriented Pig Production

This chapter briefly describes the sampling design of the household survey used in the study. Using the sample of pig-producing and nonpig-producing households, a market participation model is estimated to explain why some smallholders engage in pig production whereas others do not. The chapter also focuses on identifying factors that significantly increase the level of participation in pig production by households.

Study Areas

Southern Tagalog (Region 4) is the largest pig-producing region in the Philippines. The area has witnessed a rapid expansion in commercial pig farms in the past decade, mainly concentrated in the CALABARZON subregion. At the same time, smallholder pig producers compete for market share with larger farms. In the CALABARZON area, there are three major pig-producing provinces: Batangas, Laguna, and Rizal, whose combined pig inventories account for about 80 percent of the subregion's total (Table 3.1). Although smallholder pig production has been virtually wiped out in the province of Rizal, it still exists in the provinces of Batangas and Laguna. These two provinces represent an area where smallholder pig producers dynamically compete for market share with larger commercial farms. Thus, these two provinces were chosen as the venue for the study. Specifically, three cities that were among the top-ranked towns in terms of hog production in their respective provinces were chosen as survey areas: Lipa and Batangas in Batangas, and San Pablo in Laguna.

Moreover, smallholder contract producers were found only in Batangas city, and all belong to one contract farming scheme: the Sorosoro Ibaba Development Cooperative (SIDC). A sample was taken from among these hog growers under contract to make a comparative analysis between smallholder contract growers and independent hog producers (discussed in Chapter 4).

Smallholder pig production is undertaken in the *barangays* (villages) in the periphery of the city proper, in locations that would be classified as either peri-urban or rural. Lipa, Batangas, and San Pablo are classified as cities on the basis of their populations and business establishments. However, about three-quarters or more of the land area in these cities is classified as agricultural (in terms of land use) and rural (in terms of social and infrastructural facilities) (Office of the City Planning Office, San Pablo City 1996; Office of the City Planning and Development Coordinator, Batangas City 2001; Office of the City Planning and Development Coordinator, Lipa City 2001).

Although Lipa and San Pablo are situated in two different provinces south of Metropolitan Manila, they are about equidistant from the metropolis (Figure 3.1). Batangas city and Lipa

Table 3.1 Top three pig-producing provinces in the CALABARZON subregion and the share of backyard pig inventories, 2000

Subregion/ province	Total inventory (head)	Share of pig inventory (%)	Share of inventory in backyard farms (%)
CALABARZON[a]	1,263,860	100.0	35.1
Batangas	517,910	41.0	31.1
Laguna	245,540	19.4	46.8
Rizal	229,800	18.2	0.9

Source: BAS 2002.

[a]CALABARZON indicates the Cavite, Laguna, Batangas, Rizal, and Quezon provinces of Southern Tagalog.

are located in the same province, but Batangas city is farther away from Manila, some 27 kilometers south of Lipa. Well-paved major networks of the national highway system connect these three cities to one another as well as to Metropolitan Manila.

Criterion for Sampling Pig-Producing and Nonparticipating Households

To obtain samples of smallholder pig-producing households, the study initially used the definition of the Bureau of Agricultural Statistics for a backyard[8] farm in various villages of the cities chosen as sample sites. In the implementation of the field survey proper, however, the study did not restrict itself to households meeting this definition, but rather put greater emphasis on the criterion of pig production operations being household-based—that is, employing mainly the household's resources of family labor, land, and capital. Thus, households that were raising more than 20 head of pigs at the time of the survey were also included in the sample.

In Lipa and San Pablo, the sampled villages were those where households engaging in pig production were prevalent. In the city of Batangas, the sampled village was Sorosoro Ibaba, because this village had a concentration of smallholder contract farmers in pig production, in addition to being one of the major pig-producing villages in Batangas. Prior reconnaissance surveys indicated that smallholder contract farming was not practiced in any other towns in Batangas or Laguna. All smallholder contract hog farmers belong to SIDC. Approximately 50 pig-producing households ("participants") were randomly chosen from each of these three cities. An equal number of households not raising pigs or any livestock ("nonparticipants") were drawn randomly from the same sites. Table 3.2 shows the distribution of sample of participants and nonparticipants used in testing the hypotheses of this study. The actual number of samples is less than what was targeted because of insufficient and unreliable information from 4 percent of pig-producing households and 6 percent of nonparticipants.

Characteristics of Household Head

Table 3.3 compares the household characteristics of participants and nonparticipants. Household heads for both participants and nonparticipants are generally male. On average, the household heads of participants and nonparticipants are similar in age, years

[8]The Philippines Bureau of Agricultural Statistics defines a backyard farm as one holding not more than 20 head of pigs in adult-equivalent.

Figure 3.1 Locations of San Pablo, Lipa, and Batangas City relative to Metropolitan Manila

Source: http//:www.multimap.com.

of schooling, and years of residence in the village.

A higher percentage of household heads of participants than nonparticipants reported having a main occupation. Among heads of participating households, although less than half regarded hog production as their main occupation, the activity is still most often

Table 3.2 Distribution of sample households of participants and nonparticipants, 2000–2001 (percent)

Type of household	Location			All locations
	San Pablo	Lipa	Sorosoro	
Participants	48	46	50	144
Nonparticipants	50	50	41	141
Total	98	96	91	285

Source: UPLB-IFPRI-SLP 2000–2001.

cited (39 percent) as the main source of income compared to other economic activities. For nonparticipants, the dominant occupations of household heads were employment in the public or private sector (35 percent) and engagement in own business (33 percent).

For engagement in other economic activities, a greater proportion of the hog raisers (39 percent) had another source of income compared to nonparticipants (21 percent), in addition to the main occupation cited. This difference may indicate the need for additional sources of income for households engaged in pig production on the one hand, and the relative stability of the main income sources of nonparticipants, on the other.

Characteristics of Spouse of Household Head

Table 3.3 also compares the characteristics of spouses of household heads of participant and nonparticipant households. Spouses among nonparticipants had slightly more years of schooling than did spouses of participating households. A greater proportion of spouses among participants were engaged in a main occupation (57 percent) compared to their nonparticipating counterparts (45 percent). For those with a main occupation, pig raising was a dominant source of income (24 percent). A significantly greater proportion of spouses among nonparticipating households was formally employed in the government or private sectors (19 percent) compared to their participating counterparts (10 percent).

The configuration of occupational engagements of participants shows that hog raising occupies a dominant place among the economic activities that they engage in. Hog production did not constitute an absolute majority of main occupations among household heads or their spouses, but when taken together as one household unit, most participant households engage in hog production as their main occupation. If the head was engaged in another economic activity as the main occupation, it is most likely that the spouse takes over the hog raising business as her main economic engagement. This fact underscores the importance of the economic contribution of the spouse among the pig-producing households.

Engagement in supplementary economic activities appears to be more crucial for participants than for nonparticipants. Diversification could be a strategy to stabilize the flow of income associated with the business cycle in hog raising, and to support the hog-raising activity according to the flow of income and expenditures.

Resources of the Household

The main resources considered for comparison were household size, ownership or the operatorship of agricultural land, remittances from abroad, and connection to piped-in water. Average household size (defined as number of household members who are 15 years old and older), rounded off to the near-

Table 3.3 Household characteristics, participants and nonparticipants, 2000–2001

Household characteristics	Participants (n = 144)	Nonparticipants (n = 141)	Standard error (t-test)	Statistical level of significance
Household head				
Age (years)	48	47	1.71	n.s.
Male (%)	85	74	0.05	**
Schooling (years)	8	8	0.44	n.s.
Residence (years)	36	36	2.52	n.s.
Has main occupation (%)	92	80	0.04	***
Type of main occupation (%)				
Government or private employee	9	35	0.05	***
Agriculture	17	13	0.03	**
Pig raising	39	—		
Own business	7	33	0.05	***
Other	19	0	0.03	***
Existence of other/additional occupation (%)	39	21	0.05	***
Spouse of household head				
Age (years)	44	42	1.77	n.s.
Schooling (years)	8	9	0.46	*
Has main occupation (%)	57	45	0.07	*
Type of main occupation (%)				
Government or private employee	10	19	0.04	**
Agriculture	2	3	0.04	n.s.
Pig raising	24	—		
Own business	15	23	0.09	***
Other	7	0	0.04	***
Existence of other/additional occupation (%)	25	22	0.06	n.s.
Household resources				
Household size (number of individuals)	5	5	0.23	***
Has agricultural land (%)	31	20	0.05	**
Connected to piped-in water (%)	89	84	0.04	n.s.
Receives remittances from abroad	10	7	0.03	n.s.

Source: UPLB-IFPRI-SLP 2000–2001.
Notes: * indicates significant at 10 percent level; ** significant at 5 percent level; *** significant at the 1 percent level; — indicates not applicable; n.s. not significant.

est year, was similar for participants and nonparticipants (although when retained as continuous variable estimates, there was a statistically significant difference, with participant household sizes being higher). Although the ownership or operatorship of agricultural land was not prevalent in both groups, the operation of agricultural land was significantly higher among participant households than among nonparticipants. In general, both groups of households had relatively high access to piped-in water supply. Remittances from household members working abroad were not common, and there was no significant difference with respect to incidence.

Determinants of Household Participation in Market-Oriented Pig Production

The above sections presented differences in household characteristics among participants and nonparticipants that could influence the household's decision to engage in pig production. Given that pig production is on average a profitable economic activity, and that smallholder hog production is growing in this region, this section attempts to answer the following questions: What makes some households take advantage of the economic opportunities while others do not? Who is likely to participate in hog raising? When they do participate, what influences their level of market participation?

The literature, as discussed in Chapter 2, confirms that transaction costs (in various forms) play a crucial role in market participation decisions, particularly for small farmers. In the context of smallholder hog production, those transaction costs that are not directly observable are likely embedded in the underlying characteristics of the households. Following the literature that these transaction costs may arise from asymmetries across market actors in access to information and assets, we also assert that such household characteristics as those described above can be used as proxies to household access to information and assets. Where asymmetric information may not be a critical constraint, transaction costs may still ensue in the form of hidden costs of contract negotiation and enforcement, and different households may have differential responses to mitigating such costs. These household characteristics are further hypothesized to be significant in influencing the likelihood and level of participation in the hog-raising business, in combination with market factors.

Table 3.4 shows estimates of the probit equation (2.3), which is also the first step of the Heckman two-step selection model described in Chapter 2. The equation identifies the critical variables that influence the household's decision whether to engage in market-oriented pig production. Table 3.5 gives the second-step results of the Heckman selection model on the determinants of the level of participation.

The following discussion proceeds along the four sets of variables classified in Chapter 2. The empirical model includes variables representing household resources and opportunity costs, market forces and opportunities, and fixed and variable transaction costs that are incurred with participation. The empirical model was estimated using Stata® version 8 (StataCorp 2003). Subsequent iterations of various sets of covariates have resulted in a parsimonious set, the choice of which was guided by a Wald test

Table 3.4 Probit results and marginal effects of participation model for the entire sample (*n* = 285)

Variable	Coefficient	Robust standard error	Marginal effects	*z*-value
Constant	–1.24	0.38		–3.24***
Household size (HHSIZE)	0.15	0.04	0.06	3.51***
Main occupation of HH head if government or private employee (HHGOVPVT)	–0.52	0.18	–0.10	–2.82***
Main occupation of spouse if government or private employee (SGOVPVT)	–0.73	0.27	–0.29	–2.64***
Gender of the household head (HHGENDER)	0.37	0.21	0.15	1.82*
Existence of another occupation of household head (HHOTHR)	0.38	0.18	0.15	2.13**
Owns/operates land (OWNALAND)	0.45	0.19	0.18	2.32**
Years in school of household head (HHEDUC)	0.008	0.02	0.003	0.38
Existence of remittances (VREMITY)	0.29	0.31	0.11	0.93
Has piped-in water supply (WATER)	0.16	0.25	0.06	0.62
Correct predictions: 50.3%				
Wald $\chi^2(9)$: 40.25***				
Log likelihood: –174.10				

Source: UPLB-IFPRI-SLP 2000–2001.

Notes: Robust standard error is calculated using the Huber/White/sandwich estimator of variance on the assumption that the error term is not identically distributed (StataCorp 2003). Dependent variable PART is the household (HH) decision variable, which takes the value of 1 if the HH participates and 0 otherwise. * indicates significant at 10 percent level; ** significant at 5 percent level; *** significant at the 1 percent level.

Table 3.5 Determinants of levels of participation using Heckman two-step selection model (n = 285)

Variable	Coefficient	Standard error	Marginal effects	z-value
Constant	8,459.27	5,472.75		1.55
Household size (HHSIZE)	154.77	288.55	127.20	0.54
Main occupation of HH head is government or private employee (HHGOVPVT)	–1,066.94	1,109.04	–865.13	–0.96
Main occupation of spouse is government or private employee (SGOVPVT)	–913.72	1,881.17	–732.70	–0.49
Price of slaughter hogs (PSLHOG)	38.17	20.56	31.37	1.86*
Price of feed (PFEED)	–544.92	389.70	–447.84	–1.40
Price of weaners (PWEAN)	–7.66	11.07	–6.30	–0.69
Market outlets are within the village (OUTLET)	–4,367.25	886.65	–3,589.22	–4.93***
Value of breeding stock (VSTOCK)	0.003	0.027	0.002	0.10
Wald $\chi^2(11)$: 89.50***				

Source: UPLB-IFPRI-SLP 2000–2001.

Notes: Dependent variable QTYKILO is the quantity of output in kilograms (includes unsold stock). * indicates significant at 10 percent level; *** significant at the 1 percent level.

and a likelihood-ratio test for exclusion restrictions and the exclusion of some variables, such as age of household head and distance to nearest town.

Based on the empirical estimates of the final set of chosen covariates, the decision to participate in market-oriented pig production is largely influenced by household resources and opportunity costs. Note that the latter, as proxied by formal employment by the household head and spouse, may constitute a fixed cost to entry in terms of time available to engage full-time in market-oriented pig production. However, once entry has been made, the decision on the level of participation (how much to produce for sale) is largely driven by market forces, specifically, the price of slaughter hogs and the access to market outlets, which could also proxy for some variable transaction costs.

The results in Table 3.4 specifically show that household resources, as embodied in the household size (consisting of working-age household members), significantly affect the decision to participate. As hypothesized, the greater the number of household members of working age, the more likely the household is to participate in the labor-intensive activity of market-oriented pig production. However, household size did not have a significant impact on the level of participation (see Table 3.5). It is possible that household labor is considered a fixed resource that is invariant to the level of production or market transactions, in contrast to hired labor, which is more likely to be sensitive to marginal increases in the level of production and transactions.

The variables representing the opportunity costs of the household for pig production are significant in determining the decision to participate, as shown in Table 3.4. As hypothesized, the employment of the household head or the spouse (or both) in the formal sector (either government or private) acted as a deterrent to participation in hog raising. The same variables, although exhibiting the hypothesized effect (that is, a negative coefficient), did not figure significantly in the household's decision on supply or level of participation (see Table 3.5).

Given data limitations, the impact of market price variables can only be investigated in the level of participation. Table 3.5 shows that only the price of the output was statistically significant, with the expected positive sign. For the prices of inputs (weaners and feed), although these exhibit the hypothesized effects (negative coefficients), they are not statistically significant. This finding may suggest that households, in deciding on the level of market engagement, are responding to market forces (prices) more as profit maximizers (that is, level of output is not predetermined) rather than as cost minimizers (level of output is determined beforehand). This observation could be subject to further inquiry for validation.

Of the factors listed in Table 3.4 that are hypothesized to affect household-specific fixed transaction costs to market entry, the gender of the household head, the ownership of agricultural land, and the existence of a secondary occupation of the household were found to be statistically significant as determinants of the decision to participate. All had the expected positive sign of their respective coefficients.

However, the variables representing the existence of remittance income from abroad and the access to piped-in water supply are not statistically significant. Existence of remittances, although providing financial capital for engaging in market-oriented pig production, may not be the main source for such needs, hence the nonsignificant impact on household participation decision. For access to piped-in water supply, the nonsignificant coefficient may be due to the lack of variation in this variable between participants and nonparticipants, as shown in Table 3.2. The variables representing investment in phones (access to telephone services) and search and negotiation cost (proxied by distance to nearest town) were excluded from the final set of covariates based on Wald and likelihood-ratio tests.

In Table 3.5, of the factors identified as influencing household-specific variable transaction costs, the market outlet type was a significant driver of the level of market participation. The value of investments in breeding stock, however, was not statistically significant in determining the level of participation. As hypothesized, the limitation of market access for output to the village-level buyers negatively affected the level of participation. Although the coefficient of the value of breeding stock had the expected positive sign, the nonsignificance of the estimate may indicate that the variable is not effective in capturing the effect of scale on the capacity to deal with the variable transaction costs.

In summary, participation or nonparticipation in the apparently lucrative economic activity of pig production by households was shown to depend on household resources (particularly family labor), the opportunity costs of these households, and other household characteristics that facilitated access to information and resources. Households that tend to participate in pig production are those with more household labor to spare and with lower opportunity costs of time. Those with access to agricultural land and to additional sources of income were also more likely to participate. Male-headed households tended to participate. The price of output and access to markets for output outside the village were found to be important determinants of the level of participation.

CHAPTER 4

Differences in Farm Structure and Performance between Independent and Contract Farming Smallholders

This chapter focuses on the differences in household characteristics of smallholder independent producers and contract farmers. The significance of these differences will be further investigated in the next chapter. The Appendix describes in more detail the contract farming arrangement between the sampled contract farmers and their cooperative.

Comparative Household Characteristics and Types and Scale of Pig Production

The sample of independent producers and contract farmers in this study are both engaged in household-based pig production. Table 4.1 shows the characteristics of the household head, the spouse of the household head, and resources of the household for independents and contract producers. One difference found between the two groups is that both the household head and the spouse among contract farming households are on average younger than their independent counterparts. In addition, a larger proportion of both the household heads and spouses among contract farms have pig production as their main occupation compared to independents. Another differencc is that the household size of contract farms is slightly larger than that of independents.

In terms of other household resources, both groups have a high incidence of piped-in water supply, although a larger proportion of contract producers had piped-in water compared to households of independents. As for years of schooling of both household head and spouse and ownership of land, both independent and contract farm households were observed to be similar.

Smallholders were also classified by type of pig-production activity. The distinctions among production activities between independent and contract producers depend on the kind of end products that the farms produce (for example, piglets vs. slaughter hogs) and on the kind of stock invested in to produce the output (for example, gilts and sows vs. weaners). On the basis of the latter, a little less than half of the smallholders are readily classified into three specialized activities: farrow-to-wean (Type 1), farrow-to-finish (Type 2), and grow-to-finish (Type 3). More than half of them combined one activity with another: farrow-to-wean and farrow-to-finish (Type 4); and grow-to-finish and farrow-to-wean/finish (Type 5).[9]

[9]A young pig that is ready to be weaned is called by various names by practitioners. These names include piglet, weanling, weaner, and feeder stock. In this report, these terms are used interchangeably. Gilts are female pigs

Table 4.1 Household characteristics, smallholder independent and contract pig producers, 2000–2001

Household characteristics	Independents (n = 94)	Contract growers (n = 50)	Standard error (t-test)	Statistical level of significance
Household head				
Married (%)	88	92	0.05	n.s.
Age (years)	50	44	2.29	***
Male (%)	81	92	0.06	**
Years of schooling (years)	8	8	0.63	n.s.
Hog business is main occupation (%)	40	64	0.09	***
Spouse of household head				
Age (years)	46	39	2.24	***
Years of schooling (years)	8	8	0.66	n.s.
Has main occupation other than livestock (%)	42	20	0.08	***
Household resources				
Household size (number of individuals)	5	6	0.33	*
Has agricultural land (%)	33	22	0.08	n.s.
Has piped-in water (%)	84	98	0.04	***

Source: UPLB-IFPRI-SLP 2000–2001.
Notes: * indicates significant at 10 percent level; ** significant at 5 percent level; *** significant at the 1 percent level; n.s. not significant.

In general, independent producers were not clustered in the same types of activity that contract producers were. As shown in Figure 4.1, the majority (56 percent) of contract producers specialized in pig fattening (Type 3), and the rest combined pig fattening with piglet production or full-cycle operations (Type 5). In contrast, the majority of independents did not engage in specialized activities, although 22 percent of them did specialize in piglet production (Type 1). Some 36 percent of them were engaged in the combined activity of piglet production and finishing of some output to slaughter hogs (Type 4); others (20 percent) combined piglet production or full-cycle operations with pig fattening (Type 5). Independent producers were rarely found to be involved in full-cycle operation (Type 2) or pig fattening (Type 3).

The sample of pig producers was also classified according to scale of operations. As shown in Table 4.2, the grouping was made by size quintiles, based on the level of output in kilograms liveweight during 16 months of operations, as reported in the field surveys, and then converted into an annualized equivalent. Output, as used in this study, is defined as all completed sales of pigs at the time of the field survey plus the remaining stock that were scheduled to be sold at the completion of their current production cycle. The weight of the stock scheduled to be sold was computed on the basis of the age of the pig and its corresponding weight. The inclusion of weight of unsold stock was

that are more than 4 months of age and are ready for breeding or may have already been bred but have not yet ever given birth. Type 1 is a production activity where sows are kept to produce piglets that are raised to weaning age, then sold as weaners. Type 2 is a production activity where sows are kept to produce piglets and fattened to be sold as slaughter hogs. Type 3 is a production activity where weaners are purchased and fattened to be sold as slaughter hogs. Type 4 is a production activity where sows are kept with some of the offspring sold at an earlier stage as weaners, while the rest are fattened to be sold as slaughter hogs. In Type 5 production, weaners are purchased for pig fattening, but sows are also kept to produce piglets to be sold as weaners or to finish into slaughter hogs. It could also be that the main engagement is in piglet production or full-cycle operation, but weaners are purchased on the side for pig fattening.

Figure 4.1 Distribution of smallholder independent and contract pig producers by type of activity, 2000–2001

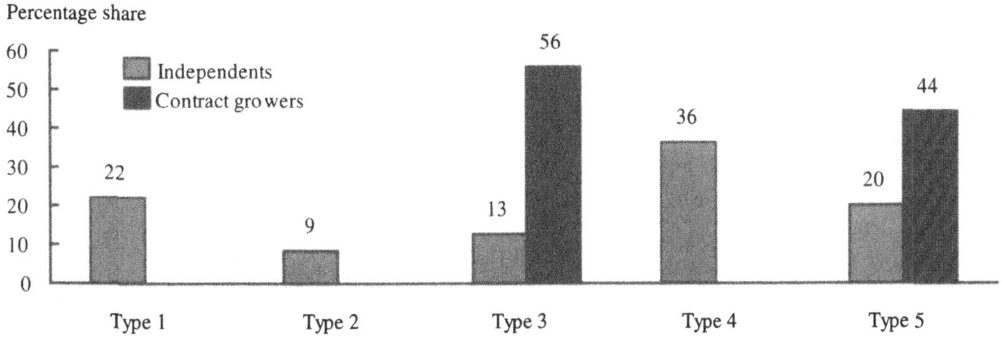

Source: UPLB-IFPRI-SLP 2000–2001.
Note: See text for definitions of Types 1–5.

made for consistency in the calculations of costs and returns to be used in Chapter 5. Note that the value of unsold stock, if ignored, would understate the value of total output over the 16-month period relative to the cost already incurred (for example, growing stock and feed).

Figure 4.2 shows a wide gap between the smallest and the largest farms in terms of the scale of activity. In quintiles, smallholders in the lowest quintile produce on average a very small volume of output (315 kilograms per year), equivalent to about 5 head of slaughter hogs (converted at 70 kilograms

Table 4.2 Distribution of smallholder independent and contract pig producers by quintile and by type of activity, 2000–2001

	Type of activity[a]					
Quintile	Type 1 (*n* = 21)	Type 2 (*n* = 8)	Type 3 (*n* = 40)	Type 4 (*n* = 34)	Type 5 (*n* = 41)	All activities (*n* = 144)
Independent farms (number)	21	8	12	34	19	94
Q1 (%)	19.1	—	4.3	1.1	6.4	30.9
Q2 (%)	3.2	4.3	2.1	14.9	6.4	30.9
Q3 (%)	—	2.1	4.3	18.1	3.2	27.7
Q4 (%)	—	2.1	2.1	2.1	4.3	10.6
Q5 (%)	—	—	—	—	—	—
All independents (%)	22.3	8.5	12.8	36.2	20.2	100
Contract farms (number)	0	0	28	0	22	50
Q1 (%)	—	—	—	—	—	—
Q2 (%)	—	—	—	—	—	—
Q3 (%)	—	—	—	—	6.0	6.0
Q4 (%)	—	—	22.0	—	16.0	38.0
Q5 (%)	—	—	34.0	—	22.0	56.0
All contracts (%)	0	0	56.0	—	44.0	100
All farms (%)	14.6	5.6	27.8	23.6	28.5	100

Source: UPLB-IFPRI-SLP 2000–2001.
Notes: The average output for each quintile is: Q1, 315 kilograms liveweight; Q2, 950 kilograms liveweight; Q3, 1,968 kilograms liveweight; Q4, 4,613 kilograms liveweight; Q5, 14,411 kilograms liveweight. — indicates not applicable.
[a]See text for the definitions of the types of activities.

Figure 4.2 Distribution of smallholder independent and contract pig producers according to size quintile, 2000–2001

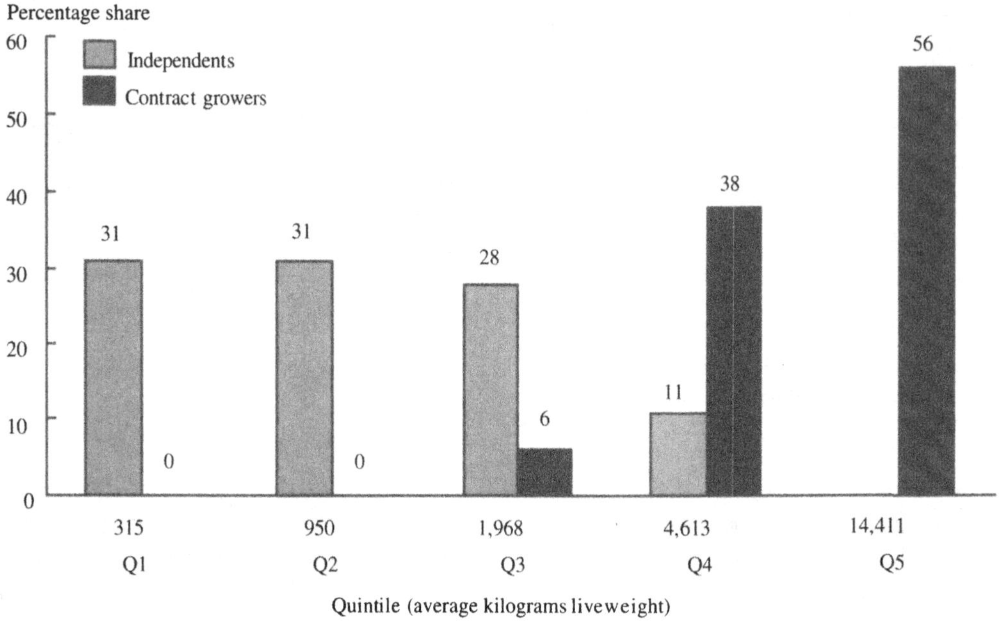

Source: UPLB-IFPRI-SLP 2000–2001.

per head), or about 20 head of piglets (at 15 kilograms per head), over the course of the entire year. On the other extreme, smallholders in the highest quintile produce more than 40 times the output of the first quintile (14,411 kilograms per year), equivalent to around 200 head of slaughter hogs per year.

Independent producers in general tend to produce lower levels of output than contract growers. Most of the independents are concentrated in the first three quintiles. Contract growers are mostly in the fourth and fifth. Among independents, most of those in the lowest quintile specialize in piglet production (Type 1), whereas those that combine piglet production with full-cycle operations (Type 4) fall in the second and third quintiles. For smallholders in piglet production, the lowest quintile average output could be achieved by holding two sows at any given time, each having a single farrowing per year, and raising to weaning age 10 piglets from each sow. For smallholders in the highest quintile, consisting entirely of contract growers, the average level of output could be achieved by the household going into intensive pig fattening (Type 3), raising around 70 fatteners per cycle, and achieving three production cycles over the course of a year. This level of output is readily achievable, as the fattening cycle spans about three or four months.

Cost Structure of Smallholder Production Activities

The three specialized activities of piglet production (Type 1), full-cycle operation (Type 2), and pig fattening (Type 3) are taken as basis for comparison of the cost structure of pig production engaged in by smallholders in the sample. The cost structure of the two other activities (Type 4 and Type 5) would be somewhere in between that of any of the two or three specialized activities that are combined.

In Table 4.3, the costs of pig production are classified into cash and noncash costs by type of activity. Cash costs mainly comprise the expenditures on intermediate inputs (mixed feeds, weaners, among others),

Table 4.3 Cost structure of smallholder pig production systems, 2000–2001

	Type of activity[a]					
	Type 1a (n = 8)		Type 2 (n = 21)		Type 3 (n = 40)	
Cost	(Php per farm)	(%)	(Php per farm)	(%)	(Php per farm)	(%)
Cash costs						
Weaners	0	0.0	0	0.0	187,707	36.7
Feeds	18,278	56.1	42,853	71.6	256,786	50.3
Transport of feeds	136	0.4	123	0.2	1,488	0.3
Veterinary medicine and drugs	223	0.7	741	1.2	3,801	0.7
Hired labor	0	0.0	0	0.0	2,336	0.5
Total cash costs	18,637	57.2	43,718	73.1	452,118	88.5
Noncash costs[b]						
Unpaid family labor[c]	8,943	27.5	10,332	17.3	7,256	1.4
Depreciation	1,544	4.7	388	0.6	2,398	0.5
Interest on operating expenses	1,864	5.7	4,372	7.3	45,212	8.9
Opportunity cost of capital	1,226	3.8	522	0.9	3,851	0.8
Opportunity cost of breeding stocks	352	1.1	505	0.8	0	0.0
Total noncash costs	13,929	42.8	16,119	26.9	58,717	11.5
Total costs	32,566	100.0	59,836	100.0	510,835	100.0

Source: UPLB-IFPRI-SLP 2000–2001.
[a]See text for the definitions of the types of activities.
[b]Excludes opportunity cost of land; opportunity cost of capital and interest on operating expenses were set at 10 percent.
[c]Combined share of operator's labor (81–87 percent) and other family labor (13–19 percent).

whereas noncash costs consist mostly of household-based resources (unpaid family labor, opportunity cost of other resources used). All entries in the cost components were translated into their relative shares to total cost. The value of total cost of the activities is expressed in local currency (PhP).[10]

For all activities, cash costs occupy the greater proportion of all costs involved. This predominance simply indicates that smallholders are highly integrated in the markets for inputs, whatever type of pig-production activity they are engaged in. The extent of their market integration does not conform to the conventional notion that backyard farms are backward in the realm of production technology, where they are associated with the use of household wastes, crop, and other farm residues as main inputs in the production process. Differences in the cost structure of the three specialized activities engaged in by smallholders lie in the intensity with which the various cash and noncash components are used.

On one extreme, the most deeply integrated in the market for intermediate inputs is the pig fattening activity (Type 3), where the share of the cash component is very high at 90 percent of total cost, with the cost of mixed feed and weaners jointly taking a disproportionately large part. At the other end of the spectrum of smallholders, those specializing in piglet production (Type 1) are the least immersed in the inputs market, where more than 40 percent of costs are in noncash form. The dominant noncash component is in unpaid family labor, making up more than a quarter of total production cost.

[10]For U.S. dollar equivalents, the conversion rate at the time of the field survey (in 2000) of PhP 50 to US$1 can be applied.

One striking difference in the cost structure of pig fattening (Type 3) from the two other activities is that the cost of weaners purchased is more than one-third of the total cost (second only to the share of feed in this activity). In the two other activities, this component is nonexistent. The absence of cost of weaners partly accounts for the apparent higher share of feed costs in Types 1 and 2 activities compared to pig fattening. The activities in piglet production and in full-cycle operation are such that the breeding stock (gilts and sows) are classified as investments, but their productivity is to be sustained by expenditures on mixed feeds throughout the production cycle. The maintenance of stock is thus partly reflected in the increased cost share of feed.

The other salient difference between the cost structure of pig fattening and the two other activities is the low share (1.5 percent) of unpaid family labor in pig fattening, which is a main household resource. For the other two activities, unpaid family labor comes out as the second most important cost component.

Finally, in terms of the total cost of production of engaging in each of the three activities, piglet production (Type 1) entailed the least total cost, and in particular, the least cash cost. For those engaged in full-cycle operation (Type 2), total costs are almost double those of the former. For those engaged in pig fattening (Type 3), total costs were on average eight times those in full-cycle production.

Thus, for smallholders to engage in piglet production with a couple of sows, the resources required are not as demanding as when they venture into the full cycle, where additional resources must also be generated to finance the cost of mixed feed to fatten the hogs up to the slaughter age. In addition, the production cycle in piglet production is much shorter than in full-cycle operations. This difference partly explains the nonproliferation of full-cycle operations by the sample smallholder independent producers, in contrast to their greater involvement in piglet production.

Engaging in pig fattening (Type 3) by smallholders is the most cash-demanding of the activities, although the production cycle is the shortest. The cost of weaners is a cash cost put up front, which is not easy for smallholders to finance on their own. Engaging in intensive pig fattening requires financing based on the production and expenditure cycle of the activity. The difficulty of financing also partly explains why engagement in pig fattening is not a popular activity among the sample of independent producers.

Comparative Access to Assets and Services

Differences in Sources and Prices of Stock

In general, the stock for engaging in pig production (weaners, gilts, or sows) is sourced either from informal markets, such as other neighboring smallholder producers, or from formal markets, such as commercial farms where more organized breeding takes place. Table 4.4 compares the sources of stock for independent and contract producers. Independent producers largely obtain piglets or weaners from neighboring backyard farms, their relatives, or their own stock. In contrast, contract producers mainly acquire their stock from their cooperative.

For breeding stock, sources of sows are similar for both independent and contract producers. Most come from neighboring backyard farms, relatives, or their own stock. Thus, in piglet production (or the finishing of some piglets to slaughter hogs), independent and contract producers behave similarly in terms of the dominant source of sows.

The prices paid by contract producers for their weaners were around 42 percent higher than those paid by independent producers on a per head basis. The difference in the prices paid was not due to the differences

Table 4.4 Sources of and prices for stock by smallholder independent versus contract pig farms, 2000–2001

	Independents (n = 94)		Contract farmers (n = 50)		All participants (n = 144)	
Source	**Number**	**(%)**	**Number**	**(%)**	**Number**	**(%)**
1. Weaners						
Neighboring backyard farm	34	77	2	5	36	44
Integrator/company/cooperative	3	7	34	92	37	46
Commercial hog farm	4	9	—	—	4	5
Special government program	1	2	—	—	1	1
Own stock	2	5	1	3	3	4
Total	44	100	37	100	81	100
2. Sows						
Neighboring backyard farm	26	58	5	63	31	58
Integrator/company/cooperative	2	4	3	38	5	9
Commercial hog farm	8	18	—	—	8	15
Special government program	1	2	—	—	1	2
Own stock	4	9	—	—	4	8
Relative	3	7	—	—	3	6
Others	1	2	—	—	1	2
Total	45	100	8	100	53	100

Average cost	Independents	Contract farmers	All participants	**Standard error (t-test)**
1. Average cost of weaners (PhP per head)				
Average	1,317	1,873	1,441	99***
Standard deviation of mean	45	89	52	
Number of farms	38	11	49	
2. Sows (PhP per head)				
Average	8,439	10,200	8,745	1,017(n.s.)
Standard deviation of mean	103	934	379	
Number of farms	38	8	46	

Source: UPLB-IFPRI-SLP 2000–2001.
Notes: *** indicates significant at the 1 percent level; n.s. not significant; — not applicable.

in weight class at the time of the purchase of the piglets.[11] With both independent and contract producers purchasing piglets at similar weight levels on average, the cost of piglets on a per head basis is higher for contract growers than for independent producers. There is no significant difference between the prices paid for sows by contract growers or by independent producers. Thus, for quite similar sources, the prices paid are also similar.

Differences in Sources, Prices, and Access to Credit for Feeds

Feeds could be obtained from retail stores in the city, from wholesale sources, or directly from feedmilling firms. Feedmilling firms, their distributors, and the retail shops

[11]The standard pricing scheme in the market for piglets or weaners is such that the first 10 kilograms were priced at a much higher rate than that charged for weight in excess of 10 kilograms. Under such a pricing system, smaller piglets cost more per kilogram than heavier ones.

Table 4.5 Sources of feed and access to feeds credit by smallholder independent and contract pig producers, 2000–2001

	Independents (n = 91)		Contract growers (n = 45)		All participants	
Source	**Number**	**(%)**	**Number**	**(%)**	**Number**	**(%)**
Integrator/cooperative	0	0	34	76	34	25
Feedmill	21	23	4	9	25	18
Wholesale	16	18	2	4	18	13
Retail	43	47	0	0	43	32
Others	11	12	5	11	16	12
Access to regular feeds credit	**(n = 92)**		**(n = 31)**		**(n = 123)**	
Yes	57	62	29	94	86	70
No	35	38	2	6	37	30

Source: UPLB-IFPRI-SLP 2000–2001.

may or may not give feeds credit to their buyers. Table 4.5 presents the sources of feed and access to feed credit by smallholder independent and contract producers.

The main source of feed differs for independents and contract producers. Independent producers get their feed primarily from retail stores. Contract growers, however, have fairly good access to feed from their cooperative, which operates its own feedmill and distribution system.

Almost all contract producers have access to feed credit, but the large majority of independent producers also indicated access to feed credit from their own sources. Notwithstanding this apparent similarity, the extent to which feed could be obtained on credit would likely be different between contract and independent producers. For contract producers, the feed requirement for the entire cycle contracted is provided on credit by their integrator. This expense is then charged back to the farmer at the end of the production cycle. For independent producers reporting access to feed credit, it is doubtful that the entire feed requirement for their operations, which entail even longer production cycles, would be provided on credit by their retail suppliers in such an inherently risky enterprise as pig production.

There are various types of feed used in pig production for the different kinds of stock and ages of pigs in the production cycle. Prices vary among feed types, depending on the constitution of the feed mix and prices of feed ingredients included. Feed types for younger stock are on average more expensive than those for older ones (Table 4.6).

Of the seven types of feed purchased by the smallholders, contract producers paid lower prices for three types—booster, prestarter, and grower feeds—compared to independents. For other types, there were no significant differences in prices paid between the two groups.

Behind these differences in feed prices paid by independent and contract growers, however, are differences in the quality of feed purchased. Each feed type is graded according to three general classes (A, B, and C), Class A having the highest quality and Class C the lowest. At present, there is no public grading or information system by which feed sold in the market is classified. The ratings of feed are therefore assumed to be associated with the reputation of feed companies and the brands they sell. The grade of feed within feed types is also indicated by its price, even for the same brand name—higher prices suggesting better quality.

Table 4.6 Prices paid for mixed feeds, by type, by smallholder independent and contract pig producers, 2000–2001

Feed type	Independents (Php per kilogram)	Contract growers (Php per kilogram)	Difference (%)	Level of significance
Booster	26.38	23.00	–12.8	**
Pre-starter	14.16	11.84	–16.4	***
Starter	12.78	12.56	–1.7	n.s.
Grower	10.35	10.04	–3.0	***
Finisher	9.29	10.69	15.1	n.s.
Breeder	9.48	9.62	1.5	n.s.
Lactating	10.30	10.42	1.2	n.s.

Source: UPLB-IFPRI-SLP 2000–2001.

Notes: ** indicates significant at 5 percent level; *** significant at the 1 percent level; n.s. not significant.

For contract growers, the quality of the feed they use is controlled by the cooperative through the operation of its own feedmill. For independents, however, it was difficult to derive indications of feed quality, given the lack of information on the brand names of feeds used. The majority either bought feeds that could not be readily classified or whose brand name could not be recalled.

Differences in Access to Animal Health Services

Access to animal health services is indicated by whether the farm was visited by veterinary personnel (veterinarian or paraveterinarian) at least once in 2000, at the time of the survey. Also indicative is whether the smallholder consulted with veterinary personnel when their pigs needed medical attention. The first indicator suggests that not only were there veterinary personnel in town, but these personnel in fact visited smallholders at least once a year. The second indicator suggests that veterinary service institutions were available for consultations. The high incidence of farmers traveling to town to consult with them suggests smallholder willingness to bear at least some of the cost of service in exchange for the benefit of keeping healthier animals. Table 4.7 compares access to animal health services by independent and contract producers.

Table 4.7 shows a relatively high access to animal health services by both groups. In fact, all contract growers were visited by veterinary personnel during the study period. The frequency of visits is also relatively high for both groups, with majority of the farms being visited more than twice a year. Less than a third of the independent producers consulted with veterinary personnel in town, whereas most smallholder contract producers did so.

Table 4.7 also indicates what type of veterinary personnel, government or private, visited the farm, and what type of veterinary personnel the farmer consulted in town. Independent smallholders were visited mainly by government personnel. For the contract growers, the source of veterinary services was mostly private. In seeking veterinary services in town, the role of the private sector significantly increases for independents, as reported by more than half of those who did seek veterinary services. Contract growers also went overwhelmingly for private veterinary services.

Given the relatively high incidence of farms being visited by veterinary personnel during the year, the differences in access to animal health services may lie in the quality

Table 4.7 Differences in access to animal health services by smallholder independent and contract pig producers, 2000–2001

	Independents (n = 94)		Contract growers (n = 50)	
Access	Number	(%)	Number	(%)
Farm visited by veterinarian/paraveterinarian	70	75	50	100
Farm not visited by veterinarian/paraveterinarian	23	25	0	0
Frequency of visit by veterinarian/paraveterinarian (in 2000)				
Once	13	23	4	12
Twice	10	18	5	15
More than twice	33	59	25	73
Total	56	100	34	100
Veterinarian/paraveterinarian visited by farmer	29	31	30	60
Type of veterinarian/paraveterinarian visited				
Government/public	22	92	1	3
Private	2	8	28	94
Both public and private	0	0	1	3

Source: UPLB-IFPRI-SLP 2000–2001.

of services that the smallholders receive. The source of the service—government or private sector—may indicate the quality of service received. There are debates on whether private veterinary services are better than public services. In Lipa and San Pablo, where the independent producers depended on public animal health services, the levels of activity in the respective local veterinary offices differed. However, this disparity may not be true in general. The quality and effectiveness of public veterinary services depend on the importance that local government puts on such services and on the animal health programs implemented for smallholders.

For contract producers, whether the source is private or public is not an issue because veterinary services are provided by the cooperative. The cooperative maintains a veterinary staff that provides the service to contract growers on a regular basis and charges the cost to the contract growing enterprise. Because it is in the interest of the cooperative to enable its contract producers to raise healthy and good quality slaughter hogs, which the cooperative will sell at the end of the production cycle, the cooperative seeks to provide good and effective veterinary services to its smallholder contract producers.

Differences in Access to Credit for Capital

Smallholder need for credit could not be directly determined from the survey data. To deal with this shortcoming, other indicators of access to credit for capital were derived from the available information. Indicators include credit-related responses to the survey questions on major constraints on growth and areas in which government assistance would be useful. Table 4.8 shows smallholder responses on the significance of credit for business expansion.

Lack of capital was the most commonly cited barrier to business expansion among independent producers (74 percent). Among contract farmers, only a minority cited this as a problem; instead they were more concerned with the incidence of diseases affecting their animals. In terms of desirable government assistance to smallholders in their pig-production enterprises, independent producers and contract growers view credit assistance as top priority.

Table 4.8 Perceptions of smallholder pig producers on constraints to expansion of pig production business and on major areas of government assistance, 2000–2001 (percent, multiple responses)

Constraint	Independents (n = 94)	Contract growers (n = 50)
Constraint to expansion		
Lack of capital	74	26
Incidence of diseases	34	80
High cost of feeds	26	12
Inadequate space	15	10
Areas where government can help		
Credit or finance	64	56
Feed subsidy	30	2
Veterinarian medicines or services subsidy	21	12
Training or seminar	17	2
Output price support	9	8

Source: UPLB-IFPRI-SLP 2000–2001.

The indicators of access to credit were also obtained indirectly. Smallholders whose loan applications were denied were identified as facing barriers to credit access. Among smallholders who did not apply for loans, those who reported excessive risk as their reason for not applying were considered to face credit barriers. A third indicator of credit access was the ability to obtain a loan at an interest rate of 10 percent or less per year. Those who were able to obtain credit at such low rates were considered as having access to subsidized credit.[12] Table 4.9 shows the comparison of attitudes and access to credit by independent and contract producers.

According to Table 4.9, one-fifth of the independent producers who applied for a loan were denied credit. However, none of the contract growers who applied were denied. On this count, those independent producers who were denied credit faced access barriers. In addition, close to 70 percent of the independent producers who did not apply for a loan cited "too high risks" as their reason for not applying. In contrast, only about one-third of the contract growers gave this reason for not borrowing. Finally, very few smallholders in either group were able to obtain subsidized loans.

Access to credit by smallholders appears to be associated with having formal institutional links in the production processes and in marketing. A more detailed discussion of credit facilities made available to contract growers by their cooperative is found in the Appendix.

Differences in Access to Markets for Output

As mentioned earlier, the main outputs of smallholders are slaughter hogs and piglets. The potential destination markets for slaughter hogs for independents and contract producers are largely similar—either pigs are shipped to the main live markets in Metropolitan Manila through a network of *viajeros* (long-distance traders) or they are brought to the city's slaughter house for meat distribution in the local retail markets. Feeder piglets of about 15 kilograms are sold to neighboring pig producers or are bought by

[12]As commercial borrowing rates were hovering around 15 percent per year at the time of the survey, a rate of 10 percent can be considered as subsidized.

Table 4.9 Differences in access to credit by smallholder independent and contract pig producers, 2000–2001

Characteristic	Independents Number (n = 90)	Independents (%)	Contract growers Number (n = 49)	Contract growers (%)
Tried to obtain credit for expansion	20	22	25	51
Did not try to obtain credit	70	78	24	49
	(n = 19)		(n = 25)	
Has been denied credit	4	21	—	—
Has not been denied credit	15	79	25	100
Has access to subsidized loan with ≤10 percent annual interest rate	5	33	5	20
	(n = 61)		(n = 24)	
Reason for not attempting to obtain credit for expansion				
Has no need for credit	16	26	16	67
Risk too high to take	42	69	8	33
Other	3	5	0	0

Source: UPLB-IFPRI-SLP 2000–2001.
Note: — indicates not applicable.

village traders who ship them elsewhere. Table 4.10 presents the market outlets for slaughter hogs and piglets and the prices of outputs received from these outlets.

For piglets, the main market outlets of independents were their neighbors and village agents and traders. In contrast, contract growers had their own cooperative as their main outlet, although they also sold to other buyers. Neighbors as customers are typically smallholder pig raisers themselves. Village traders sell piglets to pig raisers in other towns. Despite the apparent difference in the main market outlets between independents and contract growers, there are no significant differences in the prices they receive for piglets. Thus there is no institutional advantage associated with one group or the other in the pricing of own piglets.

In the case of slaughter hogs, a different configuration comes about. The main market outlets of independent smallholders were the local retail markets. A large proportion of them also sell to village agents and traders. For contract growers, the main outlets were their cooperative. Contract growers are bound by contract to deliver the finished slaughter hogs to their cooperative, which will then market them to its regular buyers.

A significant difference exists in the mean prices received for slaughter hogs. The group of contract producers received, on average, a price premium of about 11 percent over that received by their independent counterparts. The price advantage gained by contract growers is not due to observed geographical or locational advantage in terms of distances to main markets for output. In fact, the mean distance between the location of the contract growers (Sorosoro Ibaba, in Batangas) and the Metropolitan Manila market (93 kilometers) is significantly larger (with standard error of the mean difference equal to 1.1) than that between the independent producers (in Lipa and San Pablo) and Manila (81 kilometers). Further-

Table 4.10 Market outlets of piglets and slaughter hogs, and average prices received for outputs by smallholder independent and contract pig producers, 2000–2001

	Independents		Contract growers		
Characteristic	Price (Php per kilogram)	Number	Price (Php per kilogram)	Number	Standard error (*t*-test)
Market outlets for piglets					
Neighbors	89.80	22	92.00	1	
Village agents or traders	90.38	18	—	0	
Integrator or cooperative	90.67	5	87.00	12	
Local meat retailers	90.35	6	—	0	
Others	81.34	4	87.58	4	
Unknown	94.33	3	82.00	2	
All outlets	89.78	58	86.86	19	2.70 (n.s.)
Market outlets for slaughter hogs					
Neighbors	57.83	10	—	0	
Village agents or traders	58.43	19	—	0	
Integrator or cooperative	56.00	1	63.57	49	
Local meat retailers	56.68	25	—	0	
Others	61.30	3	—	0	
Unknown	58.40	9	—	0	
All outlets	57.78	67	63.57	49	0.71***

Source: UPLB-IFPRI-SLP 2000–2001.
Notes: *** indicates significant at the 1 percent level; — not applicable; n.s. not significant.

more, the contract growers were located significantly farther (with standard error of the mean difference equal to 0.86) from the nearest town (11 kilometers) than were independents (7 kilometers).

In part, the price advantage that contract growers enjoyed over their independent counterparts can be traced to their institutional linkages to markets. Whereas independents have to deal individually with village agents or the *viajeros,* contract growers have only to deal with their cooperative, a recognized institution for the supply of pigs to its regular clients in Metropolitan Manila and to its *viajero* clients. Moreover, the cooperative also controls for the quality of inputs and services that go into the production of slaughter hogs by their contract growers. Hence it can also assure its regular clients about the health of the animals delivered and meat quality.

Overcoming Transaction Cost Barriers Faced by Smallholder Producers through Institutional Linkage

The differences in farm characteristics between independent and contract producers investigated above reveal that smallholders as a whole face a host of transaction cost barriers. Some producers have difficulty dealing with these barriers, but others are better able to overcome them. The capacity of smallholders to generate income for the household is seen to be largely dependent on the ability of these farmers to expand their scale of operations to a level that can fully exploit the household's own resources—family labor, land and space, and managerial services. In market-oriented pig production, there are three major inputs and services to which smallholders need good access: quality stocks, quality feeds, and animal health services. Smallholders also need access to more lucrative markets for their output.

For inputs, the survey results show that independent smallholders do have access to growing and breeding stocks, but the main sources are the neighboring backyard farms. These producers, however, cannot verify the quality of these stocks and thus their productive (or reproductive) potentials. Independent smallholders also do have access to feeds. Most of their sources, however, are local retail stores, and these producers have little predictive information on the quality of such feeds.

Most independent smallholders are visited by a local public veterinarian or paraveterinarian a couple of times or more in a year. It matters, however, whether the service they receive is what they really need, and whether the service is available when they need it.

For access to output markets, independent smallholders who go beyond their neighbors and local retailers to sell their pigs must rely on village agents and *viajeros*. Having small volumes to sell at irregular intervals deprives independent smallholders of bargaining power. Because traders have little information on the quality of these pigs, smallholders receive lower prices for their pigs than they otherwise would.

Smallholder contract producers, in contrast, faced a different market environment, mediated by the contracting institution. This institution was a multipurpose feedmilling cooperative that also produced weaners from its own pig multiplier farm and engaged in the supply of live pigs for the market.

Thus there are advantages to the smallholder's engaging in a production and marketing contract. The contract grower is guaranteed provision, on credit, of the growing stock and feed consistent with the size of stock stipulated in the contract.

For stocks, the contract grower is guaranteed a good quality growing stock for each batch of weaners to fatten. The production contract specifies where the growing stock must be acquired, normally from known commercial farms or (more recently) from the cooperative's pig multiplier farm. One indication of the better genetic quality of the stock is the significantly higher average price per kilogram of weaners paid by the contract growers than those paid by independent producers for weaners of about the same weight. A further indication of the superior quality of the stock obtained from the cooperative's pig multiplier farm are the additional conditions specified by the cooperative on the contract grower on improvements that the farmer must undertake on his pig pens and other production facilities. These additional stipulations are discussed in more detail in the Appendix.

The contract specifies that the farmer use only those feeds provided by the cooperative. The cooperative determines the quality of feed that it manufactures and provides to its members and contract growers. These feeds are provided to the contract growers on credit basis. Moreover, for certain feeds crucial in pig fattening, the contract growers paid significantly lower prices than those paid by independent producers for the same types of feed at the retail level. The guarantee that the feeds used by the contract growers were of high quality, the lower prices paid for them, and the opportunity to get them on credit are significant advantages.

Another advantage enjoyed by contract growers is access to veterinary and other animal health services and supplies provided by the cooperative. Good quality animal health services are ensured, because it is in the interest of the cooperative to minimize pig mortalities (and thus enterprise losses) caused by diseases in the contract growing business. Also, it is in the interest of the cooperative to maintain the quality of output from the contracted farm to the satisfaction of its regular buyers in the live hogs and meat cuts markets.

In addition, engagement in the production contract by the smallholder allows access to more stable output markets. That the cooperative, as a reputable institution, does the marketing for the contract smallholder gives the latter an advantage over the independent small producer who would have difficulty establishing his reputation individually. The cooperative maintains a set of regular clients that includes institutional and noninstitutional outlets in the main market of Metropolitan Manila, as well as regular *viajeros*. The contract growers, mediated by the cooperative as a commercial institution in the supply of live pigs, obtain premium prices for slaughter hogs coursed through its marketing mechanism. The cooperative also maintains a reputation of supplying high quality pigs, as well as meat cuts from its processing facilities and distribution outlets. The output market advantage of the contract growers is reflected in the significantly better prices for live slaughter hogs they received, which was on average 11 percent higher than those obtained by the independent smallholders in the same region.

Finally, engagement by smallholders in the contract allowed for scale expansion. The initial scale, from 20 to 80 head of fatteners, already embodies the opportunity to expand scale of operations well beyond the average holdings of independent producers (13 head at any given time). Furthermore, a successful performance in one cycle, in terms of earning good profits for the operation, enhances the opportunity to undertake a contract for second and third cycles in a single year. The opportunities to work on a larger volume of pigs each cycle and to successfully grow pigs over two or three cycles in a single year account for the contract growers' much higher levels of output on an annualized basis than those attained by independent smallholder producers. The income potentials are substantial for smallholder households from this more intensive use of resources in pig production.

CHAPTER 5

Farm-Specific Differences in Profit Efficiency and Their Sources

Small-scale pig producers compete alongside larger-scale farms in the market for live hogs. In the wider context, backyard pig producers compete with more sophisticated large commercial farms in supplying live pigs to the main market centers. Being competitive means that one should be able to maintain or improve one's market position.

As discussed in Chapter 1, the basic yardstick of competitiveness for smallholders is that they should both earn higher unit profits (at least when family labor is not costed) and be more efficient in the use of resources per unit of output than large farms. Therefore, the comparison of farm performance in the following sections is made across the combined sample of small and medium-sized pig producers, grouped by size quintiles according to their output level.

In the next sections, a sample of specialized weaner piglet producers (Type 1 farms; see Chapter 4 for the definitions of the types of farms) are excluded from the comparisons and analysis, as the empirical focus is on all farms that produced and sold slaughter hogs. Piglet production involves distinct output not comparable to that for slaughter hogs. The market for weaners is also different from that for slaughter hogs, and piglets systematically fetch higher prices than slaughter hogs on a per kilogram liveweight basis.

As explained in Chapter 4, there are two farm types that produce weaners for sale in addition to their main activity of feeding slaughter hogs (Types 4 and 5). Farms of these types are included in the main analysis, exemplifying farms that exhibited some degree of flexibility in their operations compared to the two types of specialized farms—that is, full-cycle operations (Type 2) and pig fattening (Type 3). It is inevitable that a degree of joint-output error stems from this procedure, but it is mitigated by a focus on overall profit efficiency without looking at technical or allocative efficiency in isolation.[13]

Farm Profits per Unit of Output

Farm profits per unit of output were evaluated with and without costing family labor. Family labor includes labor of the farm operator and that of other family members assisting in pig production. The legislated minimum wage for agricultural workers at the time of the field survey (PhP 130 per person-day in 2000) was used to represent the opportunity cost of family labor. However, this value could be an overestimate of the shadow wage rate of family labor under the prevailing conditions of households, where large amounts of female and child labor

[13]The issue is the ability of smallholders to make money efficiently from keeping pigs; some households may be more flexible in their strategies than others, which is part of the issue. Econometric fixes for the joint-output problem are difficult to implement and were judged not worth the effort for purpose of the present exercise.

used in backyard hog raising are not in fact free to do commercial labor off-farm. Using the higher rate only serves to illustrate the maximum error that could be introduced by not putting a cost on family labor.

Table 5.1 presents the mean profits per unit of output of smallholders, grouped into quintiles, according to their levels of output, with and without imputing the cost of family labor. Separate entries are made for independent and contract producers.

Profit per Unit of Output by Quintile

Profit per kilogram of output from the smallest farms to larger ones follows the same pattern regardless of whether family labor is costed. Profit per unit of output sharply rises as one moves from the first to the second size quintile, then gradually falls in the third and the fourth, then moves up again at the last quintile. Even without costing family labor, the profit per unit of output of the pig producers in the lowest quintile is small.

The profit of the farms in the lowest quintile drastically deteriorates and virtually disappears when cost on family labor is included. The effect of costing family labor diminishes as pig producers move up the size scale, remaining significant, however, for the second and third size quintiles. Besides low relative profit efficiency, the smallest farms do not in fact satisfy the other condition for being competitive, that is, to earn at least as high a profit per unit of output as larger farms.

When family labor is not assigned a cost, smallholders in the second and third quintiles perform better than their larger counterparts. When family labor is costed, smaller farms still perform relatively well, with only those in the third quintile faring worse than the largest producers. The relatively good performance of small farms in the second and third quintiles even when family labor is costed is a significant finding. The results of the profit calculations tend to be underestimated when the full legislated wage rate is used to cost the labor supplied by these small farm households. Thus, for these groups of smallholders, smaller farms earn a higher profit per unit of output than larger ones, satisfying the first condition for competing in the same market with larger ones.

Independent versus Contract Farms

Mean profits per unit of output for independent and contract farms are shown in Table 5.2. When no costs were imputed to family labor, the advantage was not statistically significant, even though the independents had higher mean unit profits than did contract farms. On average, independent pig producers performed as well as contract growers.

When the cost of family labor was added, however, the mean profit per unit of output of independent producers fell significantly, but those of contract producers barely changed. The effect of charging a cost to family labor is highly significant for independent pig producers (who, in the sample, are mostly small scale), reversing the relative positions of the two groups so that the contract producers appear to be significantly more profitable. Thus the profit performance of independent producers is more sensitive to whether family labor is costed and at what level. This result seems to be driven partly by the differences in average scale, where contract producers have bigger operations relative to independent producers.

Type 1 Farms (Piglet Producers)

The group of farms engaged in Type 1 activity was segregated from the above comparisons because their output (piglets only) is distinct. Their mean profits per unit of output, with and without the cost of family labor, are shown in Table 5.3.

When family labor is not costed, mean profits per unit (on a liveweight basis) of weaner piglet producers were quite high. This partly reflects the higher market prices per kilogram liveweight for weaners than for slaughter hogs. When family labor is costed according to the legislated wage rate, the effect is drastic, turning the activity into a

Table 5.1 Profit per kilogram of output by size quintile, with and without family labor costed, independent and contract producers, annualized

Quintile	Independents (n = 73)			Contract growers (n = 50)			Standard error (t-test)		All farms (n = 123)			
	Profit without costing family labor	Profit with family labor costed	Number of farms	Profit without costing family labor	Profit with family labor costed	Number of farms	Profit without costing family labor	Profit with family labor costed	Profit without costing family labor	Profit with family labor costed	Number of farms	Standard error (t-test)
Q1	13.7	0.7	25						13.7	0.7	25	11.28 (n.s.)
Q2	21.2	16.8	24						21.2	16.8	24	3.77**
Q3	18.1	13	20	23.2	21.7	4	4.31 (n.s.)	5.09 (n.s.)	19	14.5	24	2.19 (n.s.)
Q4	15.3	11.5	4	11.3	10.6	21	3.36 (n.s.)	4.15 (n.s.)	12	10.7	25	2.48 (n.s.)
Q5				16.2	15.9	24			16.2	15.9	24	1.72 (n.s.)
All	17.5	10.0	73	15.0	14.4	50			16.5	11.8	123	

Source: UPLB-IFPRI-SLP 2000–2001.

Notes: The average output for each quintile is: Q1, 597 kilograms liveweight; Q2, 1,352 kilograms liveweight; Q3, 2,543 kilograms liveweight; Q4, 5,840 kilograms liveweight; Q5, 15,315 kilograms liveweight. The average output of all farms excluding those under Type 1 activity is 5,068 kilograms liveweight. ** indicates significant at 5 percent level; n.s. not significant.

Table 5.2 Mean farm profit per kilogram of output (Types 2–5), with and without family labor costed, independent and contract farms, annualized

Average profit	Independent farms (n = 73)	Contract farms (n = 50)	Standard error (t-test)
Without costing family labor	17.5	15.0	1.7 (n.s.)
Costing family labor	10.0	14.4	2.5*
Standard error (t-test)	2.6***	1.6	

Source: UPLB-IFPRI-SLP 2000–2001.

Notes: * indicates significant at 10 percent level; *** significant at the 1 percent level; n.s. not significant.

losing proposition. The impact of the cost of family labor on profit per unit of output is statistically significant. This result is ascribable to the intensive role of family labor among households engaged in piglet production; the imputed cost of family labor in this activity accounted for more than a quarter of the total cost of production.

All Type 1 farmers are independent pig producers. Among independents, piglet production is the second most preferred activity (second only to combining piglet production with full cycle operations). It also required the least total cost (cash and noncash) in farm operations on an annualized basis. Under conditions of low alternative employment opportunities, the relatively high profits per kilogram liveweight of output may partly explain the popularity of this undertaking among labor-abundant and resource-poor independent smallholders.

Table 5.3 Mean farm profit per kilogram of output (Type 1, n = 21 farms), with and without family labor costed, independent and contract farms, annualized

Method	Average profit (Php per kilogram)
Without costing family labor	23.8
Costing family labor	–2.0
Std error (t-test)	7.7***

Source: UPLB-IFPRI-SLP 2000–2001.

Notes: Average farm output is 367 kg. *** indicates significant at the 1 percent level.

Results of the Frontier Estimation

Farm-Specific Efficiency

As discussed in Chapter 1, there are two conditions that must hold for small farms to be competitive against larger farms. First, they must be at least as profit efficient as larger farms in the use of resources at their disposal, given prevailing prices for variable inputs and outputs, to avoid having the larger farms drive them out of business. Second, they need to earn a high enough average profit to be able to live off a small volume of output.

As discussed in the methodological section (Chapter 2), the measurement of maximum relative profit efficiency in a sample can be implemented by means of a stochastic profit frontier. The latter allows for some farms to lie above the estimated "true" frontier because of random measurement and sampling error in the econometric estimation of the frontier itself.

The dependent variable for frontier estimation is profit per unit of output, following the usual practice for normalization (Ali and Flinn 1989). The explanatory variables are farm-specific output prices (slaughter hogs), farm-specific input prices (feed and weaners), and farm-specific fixed resources (family labor, breeding stock, buildings, and equipment). The fixity of family labor and physical capital follows the convention for peasant agriculture in developing countries faced with factor market failure (Ali and Flinn 1989; de Janvry, Fafchamps, and

Sadoulet 1991). The use of farm-specific input and output prices controls for locational and other fixed-effect price differences among farms, with all farms in the sample appropriately being considered as price-takers.

The first stage involves estimating a stochastic profit frontier using fixed factors and farm-specific input and output prices. It leads to the estimation of farm-specific measures of relative profit efficiency in securing profit per unit of output, following the approach of Battese and Coelli (1995), and the associated inefficiency estimates are also generated following the same source. The second stage involves regressing the inefficiency estimates against farm and transaction characteristics that proxy the transaction cost barriers that drive farm-specific inefficiency.

Farm-Specific Efficiency across Scale of Operations

The estimated levels of mean profit efficiency of smallholders, by size quintile, are presented in Table 5.4. The behavior of relative profit efficiency, as one goes from the smallest scale to the largest, approximately follows the pattern of profit per unit by scale of operations. The least efficient group of smallholders is the group producing the lowest volume of output (54 percent). From there, relative profit efficiency sharply rises (to 75 percent of estimated maximum profit efficiency), then gradually falls from the second to the fourth quintiles, and then improves for the group of pig raisers with the highest output.

The results confirm the likely non-competitiveness of smallholders in the lowest quintile (roughly fewer than 10 slaughter hogs in a single batch per year). The results also confirm that apart from this group, smaller producers were not on average less efficient in the use of resources in generating profits, given the prices they face. Given that these mid-range smaller producers did not earn lesser profits per unit of output than did larger farms, and that they were not less profit efficient than larger farms, it is plausible that these smaller farms are able to remain in business for some time to come.

Mean relative profit efficiency was also compared between independent producers and contract growers. There was no significant difference between the level of average profit efficiency of the independent producers and that of the contract growers. The latter are in fact quite variable among themselves, with farm-specific relative profitability ranging from 60 to 84 percent of maximum.

Table 5.4 Mean profit efficiency estimates by size quintile, independent and contract farms

Quintile	Independent farms (n = 69) (%)	Contract farms (n = 48) (%)	All (n = 117) (%)	Standard error (t-test)
Q1	54	—	54	
Q2	75	—	75	
Q3	71	84	72	
Q4	61	60	60	
Q5	—	73	73	
All	69	69	69	0.04 (n.s.)

Source: UPLB-IFPRI-SLP 2000–2001.

Notes: See Table 5.1 for definitions of the quintiles. Quintiles are calculated *ex post* based on ascending farm income per batch from liveweight sales. Figures shown are means per quintile of farm-specific measures of relative profit efficiency calculated across the 117 farm sample and are consequently comparable to one another. — indicates not applicable; n.s. not significant.

Table 5.5 Maximum likelihood estimates of determinants of farm profit per unit of output, 2001

Variable	Coefficient	Standard error	*t*-value	Level of significance
Constant	35.91	3.20	11.22	***
Price of slaughter hogs (PSLHOG) (PhP per kilogram)	1.13	0.41	2.74	***
Price of feed (PFEED) (PhP per kilogram)	–16.31	1.52	–10.72	***
Price of weaners (PWEAN) (PhP per kilogram)	–8.90	0.97	–9.20	***
Family labor (FLABHRS) (hours per day)	–0.02	0.01	–1.85	*
Value of breeding stock (VBRQ) (PhP)	0.95	0.49	1.93	*
Value of building and equipment (VBEQ) (PhP)	1.70	0.87	1.95	*
Price of weaner × price of feed (PhP2)	7.87	0.84	9.42	***
Value of breeding stock × value of building and equipment (PhP2)	0.02	0.02	1.06	n.s.
Value of breeding stock × price of feed (PhP2)	–0.75	0.41	–1.85	*
Value of building and equipment × price of feed (PhP2)	–1.42	0.74	–1.93	*
Types 4 and 5 activity (COMBINE)	–0.09	0.14	–0.66	n.s.
Sorosoro grower dummy (1 if Sorosoro; SORO)	–0.31	0.14	–2.16	**
Feed conversion ratio (FCR)	–0.48	0.19	–2.48	**
Log-likelihood function: –67.1				

Source: UPLB-IFPRI-SLP 2000–2001.

Notes: Dependent variable PRFTUSQ is the profit per kilogram of output (includes unsold stock). Total number of farms included is 117 and is limited to farms with nonnegative profits and those of Types 2–5 (see text for definition of types). Six farms with negative profits were excluded to avoid introducing other biases in the data from scaling all observations to exceed 0, as required by the functional form. Price-price interactions, which were highly correlated (correlation coefficient > 0.80) with other independent variables, were dropped in the model specification. * indicates significant at 10 percent level; ** significant at the 5 percent level; *** significant at the 1 percent level; n.s. not significant.

Determinants of Farm Profit per Unit of Output

The stochastic profit frontier estimation using equation (2.4) in Chapter 2 specifies the relations among farm-specific prices and fixed factors in driving farm profit per unit of output. The results of the profit frontier estimation are presented in Table 5.5, which were estimated simultaneously with the determinants of farm-specific inefficiency in Table 5.6.

The results show that profit per unit of output was significantly and positively influenced by prices received for output and negatively by the prices paid for inputs, as expected. All fixed factors specified were significant in determining the level of profit per unit of output, and all had the expected sign, except perhaps for the small negative coefficient on family labor.

Other things equal, better prices received for slaughter hogs by smallholders were associated with higher profits per unit of output. The higher the prices smallholders paid for mixed feeds and for growing stock or weaners, the lower was their profit performance. The coefficient for the price of feed exhibited the largest magnitude among the estimated coefficients, followed by the price of growing stock. Thus profit performance was most sensitive to changes in the prices of these inputs, as expected. For production activities in pig fattening, in particular, these two cost items occupy close to 90 percent of the total cost of production.

The higher the magnitude of the material household resources represented by the value of breeding stock—as well as the value of buildings (housing) and equipment—on a per unit of output basis, the better was the profit performance. The value of the breeding stock could also represent the quality of breeding stock, where a higher value would imply higher genetic quality, keeping the number of stock constant. However, the coefficient of the number of family labor hours per day devoted to pig raising was negative and significant, although the absolute magnitude of the coefficient is very small. This trend would be predicted when labor hours are costed but is hard to understand otherwise. Perhaps higher family labor input is correlated with lower capital inputs or some other omitted factor.

Focusing on the interaction between feed price and other inputs, the interaction terms between the price of feed and that of weaners, the price of feed and the value of breeding stock, and the price of feed and the value of buildings and equipment were found to be significant. This correlation signifies not only the independent effect of the price of feed (the largest intermediate cost of production) but also its influence in combination with the other livestock production-related inputs and resources.

The coefficient of the variable representing the feed conversion ratio (FCR), computed as the quantity of feeds consumed (in kilograms) to produce a kilogram of liveweight output, turned out to be significant, with the expected negative sign. The higher the feed consumption per unit of the output produced, the lower was the profitability of the farm.

Determinants of Inefficiency: Transaction Costs and Distortions

Identifying the important factors influencing the level of farm-specific relative profit inefficiency is accomplished in the second stage of the stochastic profit frontier regression. The results of the second stage of the procedure are shown in Table 5.6.

Following the approach of Coelli, Rao, and Battese (1998), γ,[14] the variance parameter of the inefficiency effects, illustrates the importance of the transaction costs and policy distortion variables in explaining profit per unit. The estimate of $\gamma = 0.99$ (very close to 1), with its standard error of 0.004, indicates that the inefficiency effects dominate the random noise (v) in explaining profit per unit output across farms. The null hypothesis that the inefficiency effects are stochastic (Ho: $\gamma = 0$), is rejected, which implies that one should not include the hypothesized determinants (the variables on the right-hand side of the equation) of the inefficiency (u_i) equation in the frontier itself. The null hypothesis, that Ho: $\gamma = \delta_0 + \delta_1 + \ldots = 0$, is also rejected, which implies that farm-specific inefficiency effects are present. Finally, the null hypothesis, that Ho: $\delta_0 + \delta_1 + \ldots = 0$, is likewise rejected, implying that the inefficiency effects are correctly specified.

For the individual coefficient estimates, there were only two factors that are statistically significant in determining the level of efficiency (the magnitude of inefficiency) of the farm. Negatively signed coefficients reduce farm inefficiency, whereas positively signed coefficients have the opposite effect. Access to feed of known quality (GOODFEED—see Chapter 4), for example, worked to reduce inefficiency at the farm level. This result complements the effect of the price of feed on profitability in the frontier results previously discussed. Although lower feed prices are beneficial to smallholders in terms of their direct effect on profitability, the higher their confidence on feed known quality for a given price, the more efficient is farm performance.

The second significant determinant of profit inefficiency is the existence of a secondary source of income of the household head (HHOTHR). The engagement of the household head in another income-generating activity aside from pig raising may reflect the household's access to other productive ventures that would yield resources that can contribute to the financing and management of livestock activities.

The access to animal health services, proxied by the frequency of farm visits by the veterinary personnel (VETVISIT) was

[14]A value of 0 for γ indicates that the deviations from the frontier are due entirely to noise, whereas a value of 1 indicates that all deviations are due to inefficiency (Coelli 1995).

Table 5.6 Determinants of inefficiency in smallholder hog production, 2001

Variable	Coefficient	Standard error	*t*-value	Level of significance
Constant	–8.59	6.47	–1.33	n.s.
Access to reliable feed (GOODFEED)	–6.43	4.01	–1.60	*
Number of veterinarian visits to the farm (VETVISIT)	–0.29	0.20	–1.44	n.s.
Household head has other occupation (HHOTHR)	–3.07	1.92	–1.60	*
Spouse has other main occupation besides raising livestock (SPOCC)	–2.20	1.99	–1.11	n.s.
Is mixed farmer (AGLAND)	1.67	1.13	1.47	n.s.
Receives remittances (VREMITY)	2.33	1.49	1.57	n.s.
Has piped-in water (WATER)	4.72	3.32	1.42	n.s.
Market outlet is neighbor, village trader, or agent (OUTLET)	–1.20	0.95	–1.15	n.s.
σ^2	5.58	3.61	1.54	n.s.
γ	0.99	0.004	235.14	***
Mean efficiency: 69%				
Likelihood ratio test of the one-sided error: 51.04***				

Source: UPLB-IFPRI-SLP 2000–2001.

Notes: Total number of farms included is 117 and is limited to farms with nonnegative profits and those of Types 2–5 (see notes to Table 5.5). The dependent variable is constructed from the normalized deviation of a farm from the efficiency frontier (Battese and Coelli 1995), and a negative coefficient implies a reduction in inefficiency or moving closer to the frontier. * indicates significant at 10 percent level; *** significant at the 1 percent level; n.s. not significant.

not significant, although its coefficient was negatively signed, as hypothesized.

Differential Inefficiency of Smaller and Larger Farms

To determine whether results were scale-specific, the inefficiency effects stemming from transaction cost barriers and policy distortions were again tested separately for smaller and larger farms. The smallholder sample was regrouped into the smallest 50 percent and the largest 50 percent. The stochastic profit frontier and efficiency model was run separately for each group.[15] As our interest is in the differential effects of transaction cost barriers on farm level efficiency, only the efficiency results are presented, leaving out the results for the profit frontier.

The comparison of results of the differential inefficiency effects between smaller and larger farms is shown in Table 5.7. In both groups, the estimates give $\gamma = 0.99$ (very close to 1), with the standard errors also close to 0 (0.00001 and 0.0003, respectively), indicating that in both cases, the inefficiency effects dominate the random noise (v) in explaining profit per unit output across farms for the two groups.

Comparing the distinct patterns of determinants of inefficiency between the two groups, the proxy variables for transaction cost barriers directly related to access to inputs and services turned out to be significant determinants of relative profit inefficiency among the smaller-scale pig producers, but not among their larger counterparts. These are the variables relating to access to feed of known quality (GOODFEED) and access to animal health services (VETVISIT). Access to feed of known quality had the largest significant negative effect on inefficiency. That this same variable was no longer significant among the group of larger pig producers implies that almost all larger farms have access to the same quality, branded feed; in fact, many of these relatively large-scale producers were contract growers using feed milled by the owners of the pigs. The same

[15]The separate runs imply that the relative profit efficiency measures derived are not comparable between the two groups.

Table 5.7 Differential inefficiency effects of transaction costs and policy distortions on small and large farms, 2001

	Smallest 50 percent of farms ($n = 58$) (Mean efficiency = 68 percent)			Largest 50 percent of farms ($n = 59$) (Mean efficiency = 64 percent)		
Variable	Coefficient	Standard error	Level of significance	Coefficient	Standard error	Level of significance
Constant	0.90	0.72	n.s.	1.46	0.80	**
Access to feed of known quality (GOODFEED)	–1.57	0.54	***	0.34	0.63	n.s.
Number of veterinary visits to the farm (VETVISIT)	–0.07	0.03	**	–0.009	0.05	n.s.
Spouse has other main occupation besides raising livestock (SPOCC)	–0.67	0.56	n.s.	–2.18	0.63	***
Is mixed farmer (AGLAND)	–0.32	0.78	n.s.	1.02	0.46	**
Receives remittances (VREMITY)	0.26	0.92	n.s.	0.76	0.88	n.s.
Household head has other occupation (HHOTHR)	0.12	0.41	n.s.	–0.03	0.78	n.s.
Has piped-in water (WATER)	0.49	0.75	n.s.	–2.31	0.90	**
Market outlet is neighbor, village trader, or agent (OUTLET)	–0.29	0.78	n.s.	0.87	0.87	n.s.
σ^2	0.47	0.15	***	1.58	0.24	***
γ	0.99	0.00001	***	0.99	0.0003	***
Likelihood ratio test of the one-sided error	45.60		***	38.82		***

Source: UPLB-IFPRI-SLP 2000–2001.

Notes: The dependent variable and right-hand side coefficients were estimated from separated frontiers for large and small farms. ** indicates significant at the 5 percent level; *** significant at the 1 percent level; n.s. not significant.

could be said for the access to veterinary services. That access to this service did not seem to influence efficiency among the larger farms does not imply that veterinary services were not important, but rather that larger farms appeared to have managed to systematically integrate veterinary livestock services in their production programs.

For the larger-scale producers, the factors that significantly influenced their efficiency performance as a group were the existence of a secondary income source by the spouse of the household head (SPOCC), and the access by the household to a piped-in water supply (WATER). The existence of a secondary income source is more closely related to greater flexibility in the liquidity position of the household than to increased access to inputs and services directly related to market-oriented livestock production. The significance of the coefficient of the variable related to access to piped-in water appears to suggest that a reliable supply of water for livestock production is more crucial to efficiency performance at higher scales of operation than at lower scales.

The third factor that was significant in influencing the efficiency of larger farms was that of being a mixed farmer (AGLAND), although the positive coefficient was not as hypothesized. This factor should be a matter for investigation in future research.

In sum, the results above explicitly show that transaction cost barriers that are more closely related to pig production and markets affect smaller producers more than they do larger ones in determining relative levels of efficiency. The differences in performance among smaller farms are more sensitive to access to reliable inputs and services than among larger farms. Among larger farms, the nonsignificance of the same factors that matter to smaller producers implies that the former are in a better position to deal with access barriers and policy distortions than their smaller counterparts.

The relative immunity of the larger farms in the sample to the effects of transaction costs related to livestock inputs and services, which can greatly affect smaller farms, is in part because many of the larger farms were also contract farmers. The institutional arrangements of contract farmers with their cooperative guaranteed them access to the most crucial inputs and services, thus bypassing the transaction costs involved in searching, monitoring, and enforcing the relations that govern these functions.

CHAPTER 6

Policy Conclusions

This chapter summarizes and highlights the conditions under which smallholders can reasonably be expected to compete with larger, more commercially oriented farms. The chapter also sketches possible solutions—guided by the empirical evidence presented in the preceding chapters—to avoiding the precipitate ejection of smallholders from a lucrative activity in which they can actually compete, given the access to the proper markets, information, and services.

Can Smallholders Compete against Larger Farms?

The objectives of this study were to understand the market forces, structural factors, and policies that affect the relative profit efficiency of small and large farms, and to explain why some farms are more profit efficient than others. However, the study was not able to compare how well smallholder pig producers performed against the largest commercial farms, because of the unwillingness of the latter to provide data. The study did look at how well a sample of smallholders in Southern Luzon competed against a sample of medium-scale, commercially oriented specialized hog farms in the same region. As discussed earlier, competitiveness as used here refers to the capacity to generate positive profits from pig raising over time at a level high enough to derive a living from this enterprise. In the long run, relatively profit efficient farms survive, driving inefficient ones out of the market by undercutting their unit profit levels. Furthermore, by our competitiveness yardstick, smallholders must also be able to live off a small volume of output. Not only must they be relatively efficient, but they must also be able to generate a big enough total profit to survive on.[16]

Small farms are not uniform in their capacity to compete. With an average output per batch of fewer than 10 slaughter hogs and only one batch per year, the one-fifth of pig raisers in the sample that produced the smallest output were not competitive compared to even slightly larger farms. They earned the lowest profits per unit of output and were least efficient at generating profits from a given kit of resources. Hence, they appear to be the least able to survive on their pig income alone for their livelihood.

Policy to improve the incomes of this group of very small farms should be directed to helping them shift to other undertakings for which they have the potential to earn a sustainable livelihood given their resources and capacities. It is unclear that greater public support to their pig raising activities would make them better competitors in the market. A first step is to look

[16]"Hobby" farms that supplement a main income source from elsewhere can subsist a long time, even if they lose money, but are not counted as being competitive in our sense unless they are both relatively profit efficient and can generate a full livelihood on their own.

more closely at the characteristics of these very small farmers, particularly their skills and learning capacities that can help them gain employment in other productive activities. Policy and institutional intervention should take the form of both enhancing the human capital of these households and improving the environment in which business enterprises can flourish.

Apart from the smallest quintile of farms by output, the other small farms in the sample are competitive against the group of larger specialized pig producers. These smaller farms did not earn lower profits per unit of output on average than did the larger producers. They were also relatively profit efficient compared with larger producers. The econometric analysis presented in the previous chapter showed that the valuation of family labor as an input was more important to the relative profit efficiency of smaller farms than of the larger ones, as would be expected when a fixed cost (family labor stock) is spread over a larger number of units. Because many of the smallholder sample were either women with children who must stay close to home most of the time or retirees from other employment, it seems likely that the opportunity cost of family labor is lower than the actual market wage rates. Nonetheless, the willingness to put in long hours on one's own small patch is a key ingredient of the relative profit efficiency of the smaller farms. Furthermore, the analysis showed that the effect of not costing family labor as an input disappeared after reaching relatively high levels of smallholder output, although these may still be considered as small operations by world standards (batches of 80 feeder hogs at a time). Even so, such producers were relatively profit efficient.

Small-scale pig producers other than the smallest backyard operations should therefore not be dismissed out-of-hand by livestock sector personnel as backward, inefficient, and uncompetitive in the market. Pig raising offers gainful employment to those smallholders who engage in it. Given access to productive resources, technology, and services, these small producers have the capacity to generate wealth on a per unit of output basis comparable to the performance of much larger farms. Policy should focus on market forces and institutions that would enhance the technical capacity of smallholder producers other than the smallest volume operators. In addition, policy should remove policy distortions that produce an unfair playing field favoring large corporate farms. Some specific elements affecting relative competitiveness are discussed next.

Better Output Prices

The markets for live pigs are relatively competitive: smallholders have alternative market outlets. Still, some groups of smallholders get significantly higher prices per kilogram of output than do others, and this disparity cannot be traced to locational differences. Getting higher output prices was instead associated with smallholders having a direct link to organizations that have a stake in both the production and marketing of output, unlike independent small pig raisers. The latter must individually and independently search for inputs and the market outlet that offers the best prices for the pigs they sell.

A promising course of action is to promote institutional arrangements linking formal organizations engaged in feed milling, piglet production, and marketing of live hogs and meat, with smallholder pig producers. These smallholders can offer their family labor, land, facilities, manure disposal capacities, and political support in exchange for the opportunity to raise pigs according to best production practices and using the integrator's feed, growing stock, and support services. The contract farmers in Sorosoro examined in Chapter 4 and the Appendix did well using this strategy and, interestingly, were paid an 11-percent liveweight price premium on average by third-party buyers compared to independent smallholders in the same region. The price differential indicated higher confidence in the quality of the contract farmers' output.

Feed Prices

The price of feed has the largest effect of all inputs on the profitability of engagement of smallholders in pig production. Mixed feed is also the single largest component of production cost in pig raising, varying in degree of importance only according to the type of activity engaged in by smallholders. Results show that for certain feeds used intensively in pig fattening, contract growers paid significantly lower prices per kilogram than did independent producers. From the nature of the contract with the multipurpose cooperative, it is unlikely that the lower prices paid were due to lower feed quality, as the cooperative had a stake in the performance of their contract growers and the profitability of the pig production activity. The lower prices paid for feeds are attributable to their direct sourcing from the feedmilling cooperative.

Another route to reducing prices paid by independents for commercially mixed feeds is through change in trade policy. Under the current trade regime, tariff rates imposed on maize—the main ingredient for domestically produced concentrate feeds—remain relatively high: 35 percent within the minimum-access-volume (MAV) for maize, as negotiated with the World Trade Organization (WTO), and 50 percent outside quota.[17] Furthermore, because the volume of imports under MAV is regularly exceeded, quota recipients, which tend to be larger operations, are automatically favored over smaller ones.

The argument for the high tariffs on maize stems from political sensitivity to the poverty-related aspects of reducing protection for corn farmers, considered to be from poor rural households. The contribution of the corn subsector to the Philippine agricultural economy, however, has gradually declined over the years, and now constitutes an average of about 5 percent of gross value added in agriculture for the years 2000–2003 (NSCB 2005).

Tariff cuts on maize would have an important redistributive effect. Many large players in the pig industry are horizontally integrated with mills for flour for human consumption and undoubtedly benefit from the much lower tariff rates (3–11 percent) imposed on grain imported for human consumption. High maize prices also benefit a relatively small group of farms in the politically sensitive parts of the Southern Philippines, especially Mindanao. Yet of the 8.6 million farms in the Philippines in 2002, approximately 43 percent were raising pigs, and these were overwhelmingly smallholder farms, especially in Southern Luzon (Census of Agriculture and Fisheries 2002; NSO 2005). Smallholders account for more than three-quarters of all pig inventories nationwide.

Furthermore, the livestock sector, led by the pig and poultry industries, now contributes close to one-quarter of the gross value added in the agricultural economy, but will need to deal with the phasing out of safeguard measures in 2008–12 negotiated under the General Agreement on Tariffs and Trade (GATT) in 1995. Any measure that would improve pig farm income through a fall in the price of the input accounting for 70 percent of the costs of pig production will boost both the wide variety of small pig farmers and the local downstream economic activities associated with their growth.

Feed Quality

One of the most consistent determinants of relative profit efficiency, and thus of profitability over time, of smallholder pig production was the use of higher quality feed. The use of good feed, most likely in conjunction with the use of higher quality animal breeds, shows up in the taste of the meat, the percent-

[17]Taken from the tariff schedule for 2000–2005, in-quota and out-quota tariff rates for maize under tariff heading 1005.90 90 (Trade Remedies Office, Department of Agriculture 2005).

age of fat, and also in better feed conversion ratios. For contract producers, access to feed of known quality was not an issue. The feedmilling cooperative provided piglets of known breeds on credit and certified feeds directly from its own feedmill. It was in the interest of the cooperative to provide high quality feed to its contract growers to generate more income for the feed miller from its own contract growing unit.

The market for feeds is one of differentiated products. Low quality feeds are identified only after discovering their effect on feed conversion ratios or growth performance of the pigs raised, not at the time of feed purchase. High variability in the actual composition of commercial feeds (even those of the same brand) and lax or no enforcement of truth-in-labeling laws, especially for cheaper feeds, are problems. If the legal enforcement of current regulations on feed quality is insufficient to bring feed manufacturers and distributors into compliance, the quality of feeds circulating in the market may be improved by establishing mandatory feed ingredient labeling that is verified through random surprise inspections.

Labeling requirements should be simple, disclosing the ingredients, their proportions, and their nutritional properties. Spot samples could be taken for testing and verification of claims on the label. Extension courses for small livestock producers could be offered on what to look for in the purchase of feed based on information provided on the label. The choice of feed will ultimately rest with the smallholders, but the proposed measures would allow them to make more informed choices.

Prices and Quality of Weaner Stock

The price of weaner stock was also shown to be a significant determinant of the profit performance of smallholder pig producers. In pig fattening, this input is the second most important cost item, although it rarely costs more than 35 percent of the final slaughter hog price, compared to 50–60 percent for feed. It is difficult to separate the issue of price from that of the quality of stock. Lower prices typically imply lower quality stock, but they could also reflect uncertainty about the quality. For any given quality of the complementary input (feed), physical productivity results for stock quality will show up in the feed conversion ratio, which was also shown to be a significant determinant of profit performance. The quality of stock is also key to the percentage of back fat (less is better), which can be hard to predict before slaughter for the mixed exotic and landrace breeds typical of the region. The positive effect of good quality breeding stock on profitability is also indirectly suggested by the positive impact of the value of breeding stock on profits: given any level of breeding stock, higher value of the stock implies higher genetic quality.

The access to good quality stock was not an issue for contract farmers in the study, as the cooperative that owned the stock had every interest in supplying the best quality available, and had access to the large breeding farm resources to obtain it. Vertical coordination is the most likely way to build market trust and recognition for animals produced by smallholders.

In contrast, independent smallholders predominantly sourced their growing and breeding stock from their neighboring backyard pig producers, with incomplete and unverifiable information on the genetic characteristics of these feeder animals. It is noteworthy that the second most popular type of activity among independent smallholders is piglet production, where gilts or sows are bred for the production and sale of weaners. As this is an activity whose profitability was seen to be very sensitive to whether unpaid family labor is costed, smallholders who mostly engage in piglet production appear to have low opportunity cost of family labor.

Getting high quality breeding stock to smallholders who specialize in the production and sale of weaners has positive multiplier effects on smallholders throughout the local area. The diffusion of high quality breeding stock from tightly controlled commercial multiplier farms to small independent piglet producers is not likely to come about on any appreciable scale through market forces (that is, commercial pig multiplier companies supplying high-priced pure or cross-breeds to small piglet producers). External institutional assistance is likely to be needed, at least until the quality of the resulting slaughter hogs is recognized in the market. Local governments' livestock development programs could perhaps advance this function within their current animal health programs, which were observed to be highly appreciated by smallholders. Once the improved characteristics of these animals are revealed in the production systems of independent smallholders, local demand may be strong enough for regular market forces to take over.

What Is the Role of Transaction Costs Barriers in Smallholder Participation and Performance?

Participation and nonparticipation in apparently lucrative pig production by households depends in part on the available household resources and the opportunity costs of family labor. Participation is also linked to fixed transaction costs related to the capacity of households to access financial resources for market-oriented pig production and to gain a strong bargaining position in market negotiations. Households that participate in pig production usually have more household labor to spare and lower opportunity costs of time. Participants tended to be those households that were male-headed, had access to agricultural land, and had additional sources of income. The level of participation was influenced by factors that linked to variable transaction costs, the most significant of which was access to markets not limited to the village.

For smallholder pig production to be profitable and a significant source of income, smallholder producers need to reach a minimum scale of the activity that allowed relative profit efficiency. Key to expansion are generation of profits from the activity for reinvestment and access to opportunities for internal and external financing. Alternately, expansion can be accomplished by links with a formal organization that enables smallholders to generate wealth from intermediate inputs that the organization provides. This capacity to expand through external assistance is exemplified by the contract growing scheme described in the Appendix. Smallholder contract producers operated at significantly higher scale of operations than did independent smallholders, but at significantly lower levels than most medium-scale farms specializing in hog fattening.

Differential Impact of Transaction Cost Barriers to Smaller and Larger Farms

The transaction costs of access to quality inputs and services related to livestock production and marketing are important when comparing the relative performance of smaller farms to one another. However, this is not the case for that half of the sample consisting of the larger farms, as shown in Chapter 5, where all farms seem to have gotten around the problems such costs bring about. In particular, the larger farms solved the problems of access to sources of feed of known quality and had a high frequency of visits by veterinary personnel to their farms. Such was not the case for all smallholders, as discussed above in the case of quality feed.

Independent smallholder producers rely heavily on the local public animal health services for veterinary care. However, these services are not always available, nor are they of the desired level of quality. Many, but not all, smallholders had paid consultations with private veterinarians in town. An issue for further consideration is whether larger farms could take care of themselves for veterinary services through the private sector.

If so, then the public resources of the local (and national) animal health services might be retargeted to smaller farms. The exception would be public animal health services for the broader public good, such as measures for the prevention of epidemic disease.

If Transaction Costs Disproportionately Affect Smaller Producers, What Can Be Done?

The case study of the contract growing scheme of the Sorosoro Ibaba Development Cooperative with smallholder contract producers (see Appendix) demonstrates that smallholder pig producers need not be constrained by transaction costs associated with being small. The scheme enables smallholders to significantly expand their scale of operations, access good quality stock and feed on a credit basis, obtain information on better technology and management in pig production, and obtain regular animal health services. Participation in the contract scheme also assured them of markets for their output and higher prices per kilogram liveweight for their pig produce than would otherwise be the case. Although the inherent risks in pig production remain and may cause some contract growers to incur losses in some cycles, the scheme shares the risks as well as the profits between integrator and grower and is profitable in general for both smallholders and the nonfarmer member investors of the cooperative.

The positive impact of the contract model that the cooperative has devised on the provision of variable inputs and services, assistance with marketing of output, and the sharing of both risks and returns is evident. Some of the return to the scheme stems from the ability to overcome transaction costs, a net social loss, which would not have been possible otherwise. Smallholders receive quality services and get substantial de facto credit and insurance. The integrator gains labor that monitors quality without being supervised; provides all environmental services at no risk to the integrator; provides the fixed overhead, water, and electricity (in some cases) at the production level; and allows investors to reap the benefits of pig production without the discomforts.

In theory, an alternative scheme by which similar results could be achieved by smallholder livestock producers is for the public sector to engage in a similar assistance to smallholders—improved household incomes of smallholder livestock producers through expansion in economic activity and market engagement. In fact, it seems likely that the public sector could achieve the same result, but it would surely require significantly large investments.

Although policies can theoretically devise instruments to deal with each of the transaction cost barriers faced by smallholders, there may be working alternatives to direct assistance to smallholders by government functionaries. Market oriented private institutions such as the Sorosoro Ibaba Development Cooperative succeed by addressing the transaction cost barriers that smallholder producers face in a manner that benefits both smallholders and the institution. Whether it is cost effective from the social standpoint depends on alternatives, which are scarce on the ground. It also depends on the share of net gains coming from alleviation of transaction costs that were net social losses versus the share that comes from the implicit government subsidy (in the form of a tax break to the cooperative). The answer to that question is the subject for another study.

APPENDIX

The Contract Growing Scheme of the Sorosoro Ibaba Development Cooperative

This appendix serves as a supplement to Chapter 4, which presented salient contrasts between independent producers and contract farmers in their access to inputs, services, and markets. This case study supplies more detail, in a more qualitative manner, on the institutional arrangements that bind the contract growers and the cooperative. The contract growing scheme with smallholder pig producers exemplifies how transaction cost barriers that typically beset small farmers can be overcome by contract arrangements beneficial to both the small farmer and the business organization. It is interesting to know whether this experience can be replicated.

The Sorosoro Ibaba Cooperative

In the provinces of Batangas and Laguna, at the time of the field survey, only one formal institution was found, after laborious inquiries, to be undertaking contract production with smallholders. This was the Sorosoro Ibaba Development Cooperative, bearing the name of the village in which its headquarters is located. Contract sizes ranged from 20 to 80 head of fatteners.[18] A case study was therefore undertaken to look more deeply into the nature of its activities to find out the factors contributing to its relative success in the business of smallholder contracts.

The following presents a brief history of the cooperative, the range of business activities it is engaged in, and trends in membership and business growth indicators. The pig contract growing program is then described, giving particular attention to the terms of the contract between the cooperative and the smallholder producer.

Beginnings of the Sorosoro Ibaba Development Cooperative

The organization started as a small farmers' association formed by a group of farmers in 1969 in the village of Sorosoro Ibaba, in Batangas City. They were mainly driven by the interest in establishing a retail store where the members could buy basic consumption goods as well as such farm inputs as feed and veterinary supplies for pig and poultry production. Within three years (1972), the cooperative organized a contract growing scheme for pig and broiler chicken

[18]At a much later period, a credit cooperative and two feedmilling firms were found to be engaged in some form of contracts with smallholder pig producers, but with more limited involvement in the pig production operations.

Table A.1 Growth in membership, assets, share of capital, net income, and average share capital per member, SIDC, 1990–2000

Year	Membership (number)	Assets (PhP millions)	Share capital (PhP millions)	Net income (PhP millions)	Average share capital per member (PhP)
1990	692	15.061	5.203	3.770	7,519
1991	700	16.366	5.446	3.967	7,780
1992	725	22.146	6.371	7.345	8,788
1993	818	22.574	7.483	5.069	9,148
1994	940	30.858	8.332	12.141	8,864
1995	1,096	35.120	11.151	5.750	10,174
1996	1,289	46.975	12.965	12.218	10,058
1997	1,515	64.508	17.127	12.020	11,305
1998	1,641	79.989	19.250	21.771	11,731
1999	1,774	89.570	25.365	24.934	14,298
2000	2,012	114.825	32.814	14.656	16,309
Annual growth rate (%)	10.8	19.8	18.5	15.8	17.3

Sources: SIDC 2001; NSCB 2002 for the implicit price indices for 1990–2000.
Note: PhP values are at 1990 constant prices.

production following the local *paiwi*[19] practice.

In 1983, the association became registered with the Ministry of Agriculture as a development cooperative, adopting its current name—the Sorosoro Ibaba Development Cooperative, otherwise known as the SIDC. Four years later (1987), the cooperative ventured into feedmilling, which has become the central business activity of the cooperative around which all other business activities are linked. Output of the feedmilling section is sold to cooperative members who are engaged in pig or poultry production, or to members who are engaged in the business of mixed feed distribution, among others.

In the 1990s, the cooperative experienced phenomenal growth. Membership tripled within 10 years. In real terms, the value of assets had a sevenfold increase. Total share increased six times over, and the average share of capital per member more than doubled. Although the net income of the cooperative exhibited annual fluctuations, average net income in the three years up to 2000 was four times higher than the average in the first three years of the 1990s. Growth of the cooperative from 1990 to 2000 is shown in Table A.1.

As of 2000, the cooperative had expanded its spectrum of business activities, which included, among others, the following activities:

1. Pig multiplier farm—a 1,500 sow-capacity facility breeding and supplying weaners for the contract growing operations and other cooperative members engaged in pig fattening;
2. Rolling meat shop—distributing choice meat cuts to institutional clients (restaurants, schools, hospitals, and local

[19]*Paiwi* refers to the traditional agreement between households in the stewardship of livestock, wherein one household entrusts to another, usually a relative by blood or affinity, the care of one of its gilts or heifers. The steward takes care of the livestock, feeds and breeds them, and is expected to make the livestock productive. In the case of pigs, as soon as the pig is bred and offspring are produced, the one who entrusted the livestock returns to pick one or two piglets of his or her choice, to bring home as soon as they are ready to be weaned from the sow.

hotels) and homeowners in subdivisions in Batangas;
3. Convenience shop in the city proper—distributing meat, grocery items, and school supplies to a wide range of customers in the city;
4. Meat stall in Alabang district in Metropolitan Manila—directly supplying the cooperative's SIDC-labeled meat cuts to customers in a high-income district in the metropolis;
5. Savings mobilization program—accepting and managing regular savings deposits and time deposits from members and associate members with excess funds, offering interest rates that were higher than the going commercial bank rates;
6. Expanded credit line—providing loans to members in need of additional capital to expand their current business operations, for pig pen construction, and up to PhP 200,000 (equivalent to US$4,000, using the 2000 conversion rate of Php 50 to US$1) for hog feed, payable in five months after the marketing of the pigs produced.

Besides engaging in business activities, the SIDC engages in social development projects in the village where it is located. It provides annual appropriations to the local (village) government to fund social and infrastructure projects for the community.

The cooperative is professionally governed by an 11-member board of directors. A small management staff of 16 members runs the business activities of the cooperative.

The SIDC *Paiwi* or Contract Growing Program

The Program

When the *paiwi* program was launched as a major activity of the cooperative, it became an integral part of the cooperative even when the feed mill operation became the dominant activity of the cooperative. The cooperative management declares that the *paiwi* serves as a community uplift program, whereby the livelihoods of the local communities area supported by the cooperative. However, the contract growers are a sure market for about 10 percent of the cooperative's feeds, with feed transactions coursed through the convenience shop on a credit basis.

Qualifications for Entry into the Program

An applicant to the contract growing program has to be a legitimate member of the cooperative. Thus, an applicant would already have a direct stake in the organization. Among the applicants, priority is given to households residing in the *barangays* (villages) closest to the SIDC headquarters, namely Sorosoro Ibaba, Sorosoro Ilaya, Sorosoro Karsada, and Tinga Itaas. The proximity of the contract growers to the cooperative allows more frequent communication between grower and the program technical support staff, and closer supervision of performance by the program coordinators.

An applicant who has previous experience in pig raising and demonstrates familiarity with the activity is preferred over inexperienced applicants. Admittance is facilitated if the applicant already has established pig pens for the activity. These pens, however, should pass the specifications and standards set by the cooperative. First-time pig raisers are also encouraged to join, provided they undergo the regular training program on pig raising conducted by the cooperative and demonstrate knowledge, aptitude, and skill in the activity afterward.

Additional Conditions for Contract Growers Raising Fatteners from the Cooperative's Pig Multiplier Farm

The pig multiplier farm (Pig Farm) is one of the relatively new ventures of the cooperative, becoming operational in 2000. The Pig

Farm is designed to be the source of piglets or weaners of the *paiwi* program, replacing purchases of weaners from commercial farms and neighboring backyard farms. As the breeding stocks are now under the control of the cooperative, weaners from the Pig Farm possess identifiable genetic characteristics.

The stock from the Pig Farm are to be treated with greater care than stock from external sources. To raise this special class of weaners, additional investments to improve the contract grower's pig pens are required. Preference is given to contract growers with good track records, to those that have established credibility and reputation with the cooperative on contract growing, and to those that have demonstrated sanitation and cleanliness in their facilities.

Gradual investment in the improvements of the housing is required. In the first year, the pig pens must have insulated ceilings for protection and troughs for feeding. There should also be isolation pens for fatteners that need special treatment. By the second year, the pens should have special drums with nozzles as sources of drinking water and medication. The pens should have iron grills for protection and have trees nearby for natural cooling. By the third year, the contract grower should be using elevated pens with proper roofing and ventilation. A fence enclosure should also have been built for protection and security.

Terms of Contract

Provision of Resources for Production. In the hog production contract, the parties share in the provision of inputs. The cooperative provides the stock (weaners), feeds, veterinary supplies, and technical services. The prices of these inputs are agreed at the start of the contract, and costs of these inputs are credited to the contract grower. At the end of the production cycle, the costs of inputs are deducted from total sales. The contract grower provides for the facilities (housing and equipment), utilities (light, power, and piped-in water), and labor. Additionally, the contract grower is responsible for environmental services, such as structures and facilities for hog waste management and disposal.

Restrictions. Contracts range from 20 to 80 fatteners per production cycle. The number of head allowed per contract is an indicator of the scale of operation that a single household can manage, using its own resource endowments, particularly land (space) and family labor. The cooperative has recently increased the maximum number of fatteners for contracts to 100 head.

The contract grower is not allowed to keep additional, similar fatteners simultaneously with those under contract with the cooperative. This restriction is stipulated to avoid the possible diversion of the cooperative's feeds to other uses. However, the cooperative allows for some flexibility for the household's engagement in the production of piglets. The presumption is that sows and piglets do not consume the same feed types nutritionally designed for the growing and finishing of slaughter hogs.

The contract grower is not allowed to sell independently any of the pigs under contract. The marketing of pigs is the sole prerogative of the cooperative.

Monitoring. The cooperative maintains personnel that manage the *paiwi* program. Specifically, they monitor proper specifications of pig pens, feed utilization, animal health and vaccination schedules, growth performance, and setting of optimal marketing date.

The management team also monitors the occurrence of animal diseases, mortalities, and their causes. The cause of death of an animal is a significant issue in the contract growing program, primarily because it is a major cost component in the pig fattening enterprise. If the pig dies from natural causes (for example, illness), the loss is treated as part of normal losses in a risky business, and the burden is shared between the cooperative and the contract grower in terms of lower profits to divide at the end of the production cycle. If the cause of death is determined to be due to the contract grower's negligence, the contract grower is made to bear the full burden of the loss, charging the cost of the animal to his own share of the profits at the end of the production cycle. The burden of such an eventuality may be deterrent enough for contract growers to be vigilant with respect to animal mortality.

Contract Enforcement. Contract violation explicitly includes (1) keeping and raising own fatteners simultaneously with fatteners under the contract agreement; (2) using the cooperative's feeds under contract for other animals outside the contract; and (3) selling pigs without asking permission from the cooperative.

The cooperative imposes penalties on contract growers for violation of contract. The penalties are graduated in severity, in accordance with the frequency of noncompliance. The gravest is expulsion from the cooperative. Such cases may be brought up to court of law for recourse.

Marketing of Output. The cooperative determines the date to sell the pigs being raised by each contract grower. The cooperative maintains about 10 regular buyers from Metropolitan Manila and other urban centers within the region. *Viajeros* (long-distance traders) are also among the regular customers. Sales are calculated in Philippine pesos per kilogram liveweight.

Settlement Terms and Profit Sharing. All expenses on material inputs (mainly, cost of stock, feed, and veterinary supplies) are deducted from the total sales at the end of the production cycle, to the satisfaction of the contract grower. The activity profit is then divided equally between the contract grower and the cooperative.

Rates of Return to Contract Growers and Investors

Returns to Contract Growers

The returns to the contract growers of the SIDC on their investments in facilities, utilities, and family labor are their 50-percent share of profit from the venture. Household net income from contract growing was found to be thrice higher than that of independent producers. The major cost items in hog fattening consist of the growing stock and mixed feeds. These two cost components together reach as high as 90 percent of total growing costs, including costs imputed to family labor. Because the contract grower does not undertake the investments in growing stocks and feeds, rates of return to own resources invested would be relatively large given that net returns to the household are three times higher than those of independent growers.

There is an additional return for contract growers from engaging in the activity. Fifty percent of the cost of feed consumed is included in the computation of the contract grower's year-end patronage refund as a member of the cooperative. The ability to receive the patronage refund is conditional on the cooperative posting, on the whole, positive profit from all business activities.

Returns to Investors

The investors in the *paiwi* program consist of the members of the cooperative. Investment capital consists of initial paid-up capital from the members and capital build-up (50 percent of interest earnings on capital and patronage refund). Accumulated capital is then reflected as share capital. As of 2000, total share capital has reached PhP 76 million, with average share capital close to PhP 38,000 (equivalent to US$760 using the 2000 conversion rate of Php 50 to US$1) per member.

The return to investors is the interest of 12 percent per year on share capital earned. This rate is higher than that offered by commercial banks of 3.75 percent per year interest on time deposits for values ranging from PhP 25,000 to PhP 50,000 for 365 days. In addition to interest earnings on share of capital, member-investors also get a share of patronage refund from the cooperative's year-end profits, prorated in the member's accumulated value of purchases from the cooperative's merchandise.

Evidently, compared to the sure proposition of a rate of return of 3.75 percent per year from time deposits, member-investors have higher earning potentials from their investments in the cooperative. Certainly the cooperative does not always post large net incomes at the end of each year, as its profits are also affected by changes in the overall economic and market environment. On average, however, the trend in the cooperative's net income in real terms has been increasing over the years.

Special Tax Status of the Cooperative

The law exempts cooperatives from the standard Corporate Income Tax levied at 32 percent of corporate profits. The tax exemption constitutes an implicit subsidy. Similar feed milling business entities that are not cooperatives are required to pay the corporate income tax.

The cooperative's management acknowledged that the implicit subsidy is significant to the cooperative's operations: the removal of such privilege would have a direct effect on the cooperative's net income available for distribution as the patronage refund to its members at the end of each year. The significance of the implicit subsidy is provided in Table A.2 in relation to the cooperative's net income from its major activities in 2000.

The cooperative earned a total of almost PhP 34 million in 2000, about 75 percent of which was generated by feed milling, its main activity. The second largest generator of profit was the convenience store, accounting for 23 percent of the total. Net income from contract growing was only about 3 percent of the total; this consisted of the 50-percent share of the cooperative in activity profits from contract growing.

If the regular corporate tax rate of 32 percent were imposed on the total net income of the cooperative, then the tax bill would have been close to PhP 11 million in 2000. This amount is the equivalent implicit subsidy that similar-sized feed milling firms do not enjoy. The magnitude of this implicit subsidy is even larger than the absolute contribution of the contract growing operations to total net income (PhP 1 million).

At its inception in 1972, the contract growing scheme was designed to be a community development program to create employment for surplus labor among the member households of the cooperative. The apparent insignificance of the contribution

Table A.2 Net income and implicit subsidy of SIDC feed milling, contract growing, and other operations, 2000

Activity/source	Net income (Php millions)	Share of net income (%)	Implicit subsidy (Php millions)
Feed milling	25.3	74.6	8.10
Minimart	7.7	22.7	2.46
Expanded credit line	1.1	3.2	0.35
Contract growing	1.0	2.9	0.32
Others	(1.2)	(3.5)	
Total net income	33.9	100.0	10.85

Source: SIDC 2001.

of contract growing to the total net income of the cooperative provides the leeway to subsidize this activity from its larger and more important profit centers, such as the feed mill.

Advantages of Contract Growing

Access of Smallholders to Investment Capital

The cooperative held in its trust the available investment capital generated from the shares of paid-up capital from its members and from the surplus generated by the organization's business activities. From the perspective of the cooperative's management, representing the interest of its member-investors, the contract growers were a good risk (given their track record). The manifestation of such an evaluation is the entrusting of an average nearly 50 head of fatteners per household and the corresponding feed requirements for a pig-fattening venture. Furthermore, the cooperative offers a loan facility, the Expanded Credit Line–Pig Pens, for expanded investment in facilities for higher-scale operations by contract growers.

Cutting Middlemen Margins through Vertical Coordination

The difference between middlemen margins for contract growers compared with those for independent producers was not directly observed. The cutting of middlemen margins in market transactions with the SIDC contract growers could only be inferred.

On a weighted-average basis for feeds relevant to grow-to-finish operations, the price per kilogram of feed was only slightly lower for contract growers than for independent producers. More importantly, however, most of the contract growers were able to obtain their feeds at wholesale level (from the cooperative). It is asserted that middlemen margins at the retail level are higher than those imposed by the cooperative on its contract growers.

Differences in margins in the purchase of growing stock are not observable, with contract producers paying relatively higher prices for weaners. Stocks purchased are of different quality classes, however; most independents obtain their stock from neighboring backyard farms whereas the cooperative obtains stock mostly from its pig multiplier farm and other commercial farms for distribution to its contract growers. For the same class of stock as that produced in the cooperative's pig multiplier farm, the margins the cooperative charges to its contract growers are assumed to be smaller than those imposed by similar commercial pig farms on independent smallholders.

In the marketing of output, the cooperative undertakes the selling function, exercising its leverage as an institution and dealing with its regular live hogs clients in the main market of Metropolitan Manila, as well as with its institutional market clients. The price received by the cooperative is the same price that it registers to compute total revenue from the contract growers' output. The proper marketing costs, however, are explicitly charged to the activity. Direct marketing by the cooperative cuts middlemen margins that would otherwise be incurred, as the cooperative mediates between the contract grower and the live market for hogs.

Cutting Transactions Costs through Improved Quality and Reliability of Output

Prices per kilogram liveweight of slaughter hogs received by contract growers were 11 percent higher than those received by independent producers. The higher prices for output reflect both the cutting of middlemen margins and the higher quality of output.

The contract growers are assured quality stock and feed from the cooperative. The cooperative is assured that its contract growers will produce characteristics desired for the final live output. The cooperative ensures, through its animal health program and veterinary services, that the final output comes from quality hogs free of diseases. With the reputation of the cooperative as an institution that supplies good quality feeds and good quality hogs and pork, transactions are facilitated between the cooperative and its regular customers. Transaction costs are reduced because the cooperative manages transactions in the final market for hogs, so that the contract growers do not have to look for their own markets.

Matching Incentives with Benefits

The pig production contract with the cooperative stipulates the cooperative's policies on the conditions under which fattening stock is to be grown. The rules are quite stringent, particularly on the use of feeds provided and the prevention of mortalities in each production cycle.

Efficient performers are rewarded with opportunities for a repeat contract. Loans are offered for the improvement and expansion of facilities and equipment to accommodate larger contracts. Contracts are discontinued for growers who perform poorly and are not able to generate profits on a consistent basis. Furthermore, penalties are imposed on growers who violate the terms of the contract. Penalties involve the loss of income from the current contract and temporary suspension from the program. Repeated violations of the agreements lead to permanent suspension from the activity or recourse to legal sanctions.

Additional incentive, especially for larger operations, includes increase in the grower's share of the end-of-year patronage refund of feed purchases (50 percent of the value of such purchases).

The bottom line is that because the contract grower and the cooperative share equally in the activity's total profits, it is in the interest of both parties to avoid losses and maximize profit per unit of output. The outcomes of the incentives and penalty schemes that operate to enhance efficiency in production benefit both parties in the pursuit of higher overall profits. In summary, the strong performance of the Sorosoro Ibaba contract growers suggests that strong advocacy by institutions and a viable package of inputs and services allow smallholders to exploit their willingness to supply family labor for less than the prevailing wage rates.

Replicability of the Contract Farming Experience Studied

Is the successful experience of the Sorosoro Ibaba contract farmers of the SIDC replicable? Despite the apparently simple structure of the organization and terms of the contract, there are some general reasons for some pessimism. First, cooperatives have a poor record in the Philippines. Most have been initiated in response to government programs for securing special credit or financial assistance for small farmers, fishers, or other producers. The experiences with failed cooperatives in the Philippines have left quite negative impressions on the potential for success of ventures organized around such schemes. Thus, the success of the Sorosoro contract growing experience must lie in the character of the contract arrangement rather than in the mere fact of being a cooperative.

The second source of pessimism on the replicability of the SIDC contract farming experience lies in the sustainability of the special tax status granted to cooperatives in the Philippines in general. The tax status of the SIDC implies a degree of subsidy that may not be easily replicable. To date, only a handful of firms in the region are known to be engaged in contract growing with smallholders in pig production. These institutions are quite new in the contract growing venture: (1) Rosario Multi-Purpose Cooperative (RMC), located in the town of Rosario in Batangas; (2) Vision 2000 Feedmills Corporation, also located in Rosario, Batangas; and (3) Siniloan Feeds Corporation (SIFECO), in the town of Siniloan in Laguna. The first is a credit and feed distribution cooperative. The other two are regular feed milling corporations. None of the three organizations enjoys the corporate tax exemption privilege, although the RMC is applying for the tax privilege by virtue of being a cooperative. All these institutions view their contract growing operations as a social or community development arm of their main operations, a means of creating livelihood opportunities for households in their respective communities.

Focus interviews with the personnel of the RMC revealed that the cooperative makes profits from interest on loans and from mark-up (2.5 percent) on feeds purchased by the contract growers from the cooperative feed distributor. Even contract growers' share of losses is added to their debt with the cooperative, which can be repaid from the profits in the next batch of contracts. Evidently, the RMC is a profit-making institution, its main instrument being credit provision. Although it is also involved in feed distribution, it has no technical expertise in pig production or pig marketing.

The case of the two other feed milling corporations is similar. The contract growing arm of the feed milling business is explicitly declared to be a community development program, not strictly a profit center. The two corporations also have no technical expertise in pig raising and slaughter hog marketing. In all three cases, the contract growing operations do not constitute vertical coordination of a market chain as is the case for the SIDC contract farming operations.

Unique Features of the SIDC

The historical accounts of the SIDC reveal that the cooperative traces its origins from and takes pride in its charismatic founder. Inspired leadership is a rare attribute in an organization and is difficult to replicate.

After three decades, the cooperative has grown to become a tightly knit organization, involving households and communities in the four villages around the SIDC headquarters. Connecting this feature to the strong performance of the contract growing program, the management possesses reliable information on and records of the contract grower's capabilities. The cooperative's extensive information and the increasing level of expertise of its contract growers through repeated contracts reinforce each other.

Another unique feature of the SIDC is that contract farming is only a small part of a profitable feed milling cooperative that has explicit social objectives, one of the expressions of which is continued support for the contract growing program. Yet the program is undertaken with such seriousness. The large investment in the 1,500-sow pig multiplier farm was made with the intention to support the entire growing-stock requirements of the contract growers not only in volume but, more importantly, in stock quality.

The contract with the city government of Batangas to manage and operate the local public slaughterhouse has provided a new and significant market outlet for contract grower's output, this time in the form of meat cuts, adding value to the output of the program. Such contracts with the local government cannot be won without the cooperative's reputation and stature in the livestock production and marketing business.

Some Reasons for Optimism about the Replicability of the SIDC Experience

The organization is professionally managed. The business activities are distinguished from its program for social services for its members and for the neighboring communities. Although the cooperative has on its agenda the provision of quality goods and services to its members, its mission statement includes being competitive with similar private enterprises. The rising net income of the cooperative is the visible result of the drive to make profit, with income benefiting the members on the basis of their level of participation as shareholders and as consumers of the organizations' goods and services.

The contract growing operation is part and parcel of a vertically integrated system, but each component has its own management team, and each component is expected to generate a profit. The program is only as good as its ability to make profit on a sustained basis.

The success of the SIDC in its contract growing business is most likely attributable to the cooperative's ability to deliver returns to its nonlivestock-farming investors (12 percent per year) and to deliver capital for smallholder contract growers while reducing risks and transaction costs for both parties through vertical integration. This is a design element that can be replicated.

There are other reasons for optimism about the replicability of the SIDC system. The social equity objective is pursued through the achievement of higher levels of efficiency and profitability of individual households. Although the contract growing program was originally designed as a community development program to provide opportunities for employment and income for members in the community, it has developed into a sophisticated system in which the goal of creating wealth has become more dominant. The level of benefits depend to a large extent on the individual household's efficiency in using the resources at hand and its ability to generate profit, which is shared with the cooperative on a 50-50 basis.

Finally, the cooperative allows flexibility in the decisions of contract growers to raise livestock not directly in competition with the activity under contract. This discretion allows the contract grower an added source of stability to household income against the inherent risks that still remain in pig fattening.

References

Ali, M., and J. C. Flinn. 1989. Profit efficiency among Basmati rice producers in Pakistan Punjab. *American Journal of Agricultural Economics* 71 (2): 303–310.

Amemiya, T. 1981. Qualitative response models: A survey. *Journal of Economic Literature* 19: 1483–1536.

Asian Development Bank. 2005. *Key indicators 2005.* <http://www.adb.org/Documents/Books/Key_Indicators/2005/default.asp>.

BAS (Bureau of Agricultural Statistics). 2001, 2002. *Inventory of hogs in backyard and commercial farms, by region, 1980–2000.* Quezon City, Philippines: BAS.

———. 2004. *Directory of commercial hog farms.* November 2004 update. Quezon City, Philippines: BAS.

———. 2005. *SWINE: Inventory in backyard and commercial farms, by province, by quarter, 2001–2003.* Quezon City, Philippines: BAS.

Battese, G. E. 1992. Frontier production functions and technical efficiency: A survey of empirical applications in agricultural economics. *Agricultural Economics* 7(3/4): 185–208.

Battese, G. E., and T. J. Coelli. 1995. A model for technical efficiency effects in a stochastic frontier production function for panel data. *Empirical Economics* 20: 325–332.

Bellemare, M. F., and C. B. Barrett. 2006. An ordered tobit model of market participation: Evidence from Kenya and Ethiopia. *American Journal of Agricultural Economics* 88 (2): 324–337.

Coase, R. H. 1960. The problem of social cost. *Journal of Law and Economics* 3 (1): 1–44.

Coelli, T. J. 1992. A computer program for frontier production function estimation. *Economic Letters* 39 (1): 29–32.

———. 1995. Estimators and hypothesis tests for a stochastic frontier function: A Monte Carlo analysis. *Journal of Productivity Analysis* 6 (6): 247–268.

Coelli, T., D. S. P. Rao, and G. E. Battese. 1998. *An introduction to efficiency analysis.* Boston: Kluwer Academic Press.

Cuevas, C. E. 1988. *Transaction costs of borrowing and credit rationing in developing countries.* Economics and Sociology Occasional Paper 1470. Columbus: Agricultural Finance Program, Department of Agricultural Economics and Rural Sociology, Ohio State University.

Cuevas, C. E., and D. H. Graham. 1986. Rationing agricultural credit in developing countries: The role and determinants of transaction costs for borrowers. In *Agriculture in a turbulent economy,* ed. A. Maunder and A. Renborg. Brookfield, Vt., U.S.A.: Grower Publishers.

de Janvry, A., M. Fafchamps, and E. Sadoulet. 1991. *Peasant household behavior with missing markets: Some paradoxes explained.* Working Paper 578. Berkeley, Calif., U.S.A.: Department of Agriculture and Resource Economics, University of California, Berkeley.

Evenson, R., and J. Roumasset. 1986. Markets, institutions and family size in rural Philippine households. *Journal of Philippine Development* 23 (1/2): 141–162.

FAO (United Nations Food and Agriculture Organization). 2005a. *FAOSTAT database.* <http://faostat.fao.org>.

———. 2005b. *Food balance sheets for 1990–2002.* <http://faostat.fao.org>.

Fenwick, L. J. 1998. Use of credit and its impact on small-scale farmer development in Kwazulu, Natal. M.S. thesis, University of Natal, Pietermaritzburg.

Fried, H. O., C. A. Knox Lovell, and S. S. Schmidt, eds. 1993. *The measurement of productive efficiency: Techniques and applications.* New York: Oxford University Press.

Goetz, S. J. 1992. A selectivity model of household food marketing behaviour in Sub-Saharan Africa. *American Journal of Agricultural Economics* 74 (May): 444–452.

Greene, W. 1997. *Econometric analysis,* 3rd ed. Upper Saddle River, N.J., U.S.A.: Prentice-Hall.

Hazell, P. B. S., and A. Roell. 1983. *Rural growth linkages: Household expenditure patterns in Malaysia and Nigeria.* IFPRI Research Report 41. Washington, D.C.: International Food Policy Research Institute.

Heckman, J. J. 1979. Sample selection bias as a specification error. *Econometrica* 47 (1): 153–161.

Heltberg, R., and F. Tarp. 2001. Agricultural supply response and poverty in Mozambique. Institute of Economics, University of Copenhagen. Paper presented at the Conference on Growth and Poverty, World Institute for Development Economic Research, 26–27 May, Helsinki.

Holloway, G., C. Nicholson, C. Delgado, S. Staal, and S. Ehui. 2000. Agro-industrialization through institutional innovation transaction costs, cooperatives and milk-market development in the East-African highlands. *Agricultural Economics* 23 (3): 279–288.

Huang, C. J., and F. S. Bagi. 1984. Technical efficiency on individual farms in northwest India. *Southern Economic Journal* 51 (1): 108–115.

Huang, C. J., A. M. Tang, and F. S. Bagi. 1986. Two views of efficiency in Indian agriculture. *Canadian Journal of Agricultural Economics* 34 (July): 209–226.

Jondrow, J., C. A. K. Lovell, I. S. Materov, and P. Schmidt. 1982. On the estimation of technical inefficiency in the stochastic frontier production function model. *Journal of Econometrics* 19 (2–3): 233–238.

Kalirajan, K. 1984. Farm-specific technical efficiencies and development policies. *Journal of Economic Studies* 11 (3): 3–13.

———. 1989. On measuring the contribution of human capital to agricultural production. *Indian Economic Review* 24 (2): 247–261.

———. 1991. The importance of efficient use in the adoption of technology: A micro panel data analysis. *Journal of Productivity Analysis* 2 (2): 113–126.

Kalirajan, K., and J. Flinn. 1983. The measurement of farm-specific technical efficiency. *Pakistan Journal of Applied Economics* 2 (2): 167–180.

Kalirajan, K., and R. T. Shand. 1986. Estimating location-specific and firm-specific technical efficiency: An analysis of Malaysian agriculture. *Journal of Economic Development* 11 (2): 147–160.

Key, N., E. Sadoulet, and A. de Janvry. 2000. Transaction costs and agricultural household supply response. *American Journal of Agricultural Economics* 82 (May): 245–259.

Kumbhakar, S. C., B. Biswas, and D. V. Bailey. 1989. A study of economic efficiency of Utah dairy farmers: A system approach. *Review of Economic Statatistics* 71 (4): 595–604.

Lanzano, L., and R. Evenson. 1997. *The effects of transaction costs on labor market participation and earnings: Evidence from rural Philippine markets.* Economic Growth Center Discussion Paper 790. New Haven, Conn., U.S.A.: Yale University.

Lapar, M. L., G. Holloway, and S. Ehui. 2003. Policy options promoting market participation among smallholder livestock producers: A case study from the Philippines. *Food Policy* 28 (3): 187–211.

Maddala, G. S. 1983. *Limited-dependent and qualitative variables in econometrics.* Cambridge: Cambridge University Press.

Ngwenya, S. A., G. E. Battese, and E. M. Fleming. 1997. The relationship between farm size and the technical inefficiency of production of wheat farmers in the Eastern Free State, Province of South Africa. *Agrekon* 36 (3): 283–301.

NSCB (National Statistical Coordination Board). 2002. *Philippine statistical yearbook.* Makati, Philippines: NSCB.

———. 2005. *Philippine statistical yearbook.* Makati, Philippines: NSCB.

NSO (National Statistics Office). 1991. *Census of agriculture and fisheries.* Manila: NSO.

———. 2002. *Census of agriculture and fisheries.* Manila: NSO.

———. 2005. *Census of agriculture and fisheries.* Manila: NSO.

Office of the City Planning and Development Coordinator, Batangas City. 2001. *Batangas City facts and figures CY 2000.* Batangas City, Philippines.

Office of the City Planning and Development Coordinator, Lipa City. 2001. *Lipa City facts and figures CY 2000.* Lipa City, Philippines.

Office of the City Planning Office, San Pablo City. 1996. *San Pablo City facts and figures CY 1996.* San Pablo City, Philippines.

Saito, K. A., and D. P. Villanueva. 1981. Transaction costs of credit to the small-scale sector in the Philippines. *Economic Development and Cultural Change* 29 (3): 631–640.

SIDC (Sorosoro Ibaba Development Cooperative). 2001. *SIDC annual report.* Sorosoro Ibaba, Batangas City, Philippines: SIDC.

Singh, I., L. Squire, and J. Strauss, eds. 1986. *Agricultural household models: Extensions, applications, and policy.* Published for the World Bank. Baltimore: Johns Hopkins University Press.

Staal, S. J., C. Delgado, and C. Nicholson. 1997. Smallholder dairy transaction costs in East Africa. *World Development* 25 (5): 779–794.

Staal, S. J., C. Delgado, I. Balteweck, and R. Kruska. 2000. Spatial aspects of producer milk price formulation in Kenya: A joint household–GIS approach. Contributed paper presented at the International Association of Agricultural Economists meeting, August 2000, Berlin.

StataCorp. 2003. *Stata statistical software.* Release 8.0. College Station, Tex., U.S.A.: Stata Corporation.

Taylor, T. G., and J. S. Shonkwiler. 1986. Alternative stochastic specifications of the frontier production function in the analysis of agricultural credit programs and technical efficiency. *Journal of Development Economics* 21 (1): 149–160.

Trade Remedies Office. 2005. Tariff schedule for 2000 to 2005. Department of Agriculture, Philippines. <http://www.da.gov.ph>.

UPLB-IFPRI-SLP (University of the Philippines Los Baños—International Food Policy Research Institute–Systemwide Livestock Program). 2000–2001. UPLB-IFPRI-SLP field survey 2000–2001. Laguna, Philippines.

Williamson, O. E. 1989. Transaction cost economics. In *Handbook of industrial organization,* vol. 1, no. 1, ed. R. Schmalensee and R. D. Willig. Amsterdam: North-Holland.

Zander, R. 1992. Limits of the transaction cost approach in the comparative analysis of financial intermediaries. Paper presented at A Seminar on Pioneer, Problems and Premises of Rural Financial Intermediaries in Developing Countries, 17–19 November, Wageningen, Netherlands.

Made in the USA
Las Vegas, NV
10 November 2026

57851003R00050